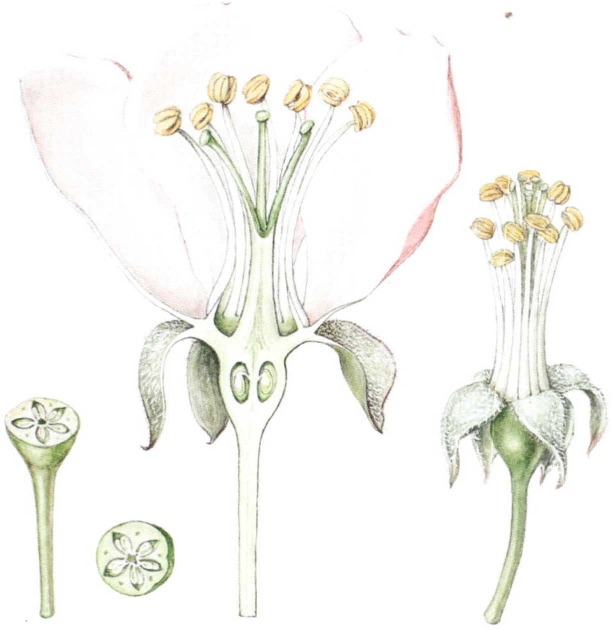

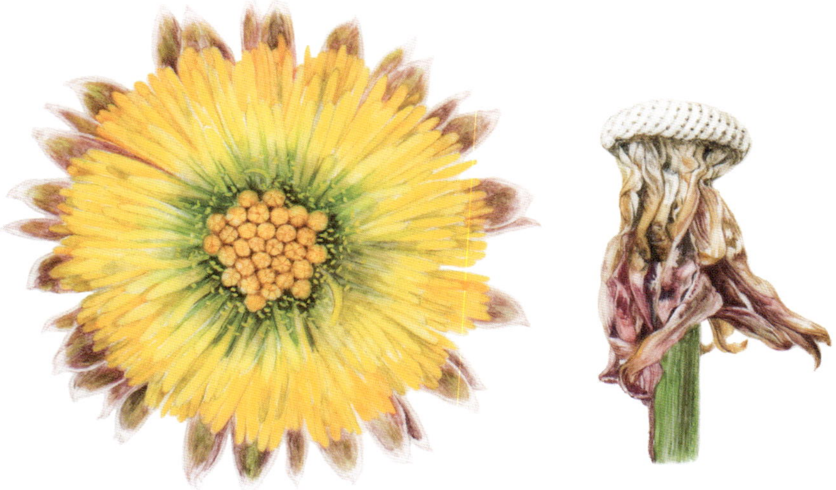

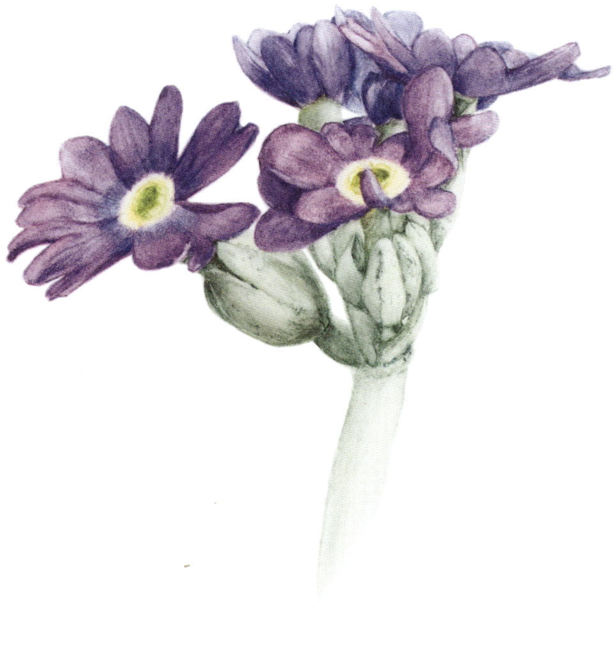

*From Trevor
September
2021*

Scottish Plant Lore

An illustrated flora

Gregory J. Kenicer

BIRLINN

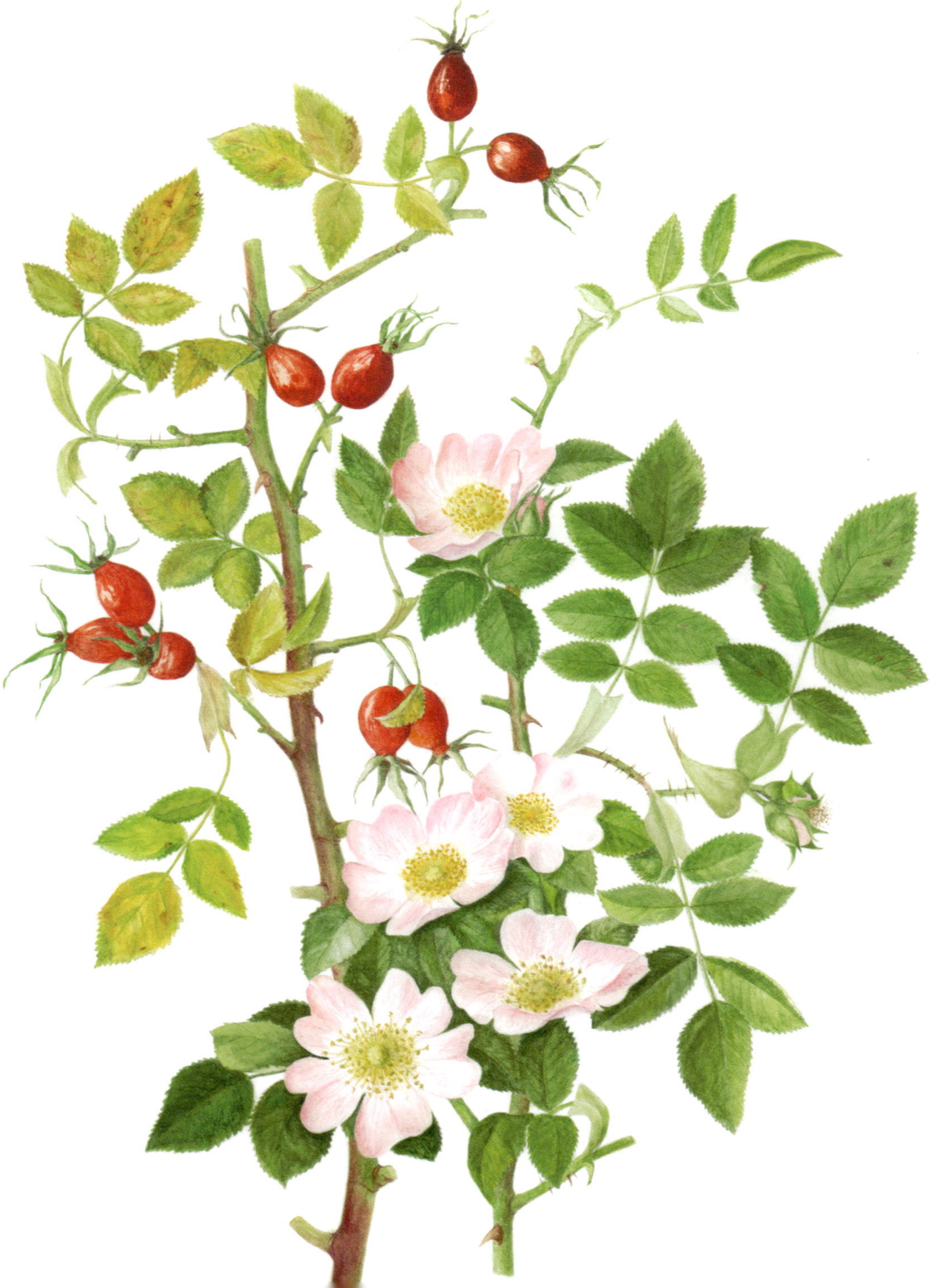

For Kim, Joe and Sam

This edition published in 2020 by
Birlinn Ltd
West Newington House
10 Newington Road
Edinburgh
EH9 1QS

www.birlinn.co.uk

ISBN: 978 1 78027 690 8

Copyright © Royal Botanic Garden Edinburgh, 2018

First published by the Royal Botanic Garden, Edinburgh in 2018
www.rgbe.org.uk

All rights reserved. No part of this publication may be reproduced, stored, or transmitted in any form, or by any means, electronic, mechanical or photocopying, recording or otherwise, without the express written permission of the publisher.

British Library Cataloguing-in-Publication Data
A catalogue record for this book is available on request from the British Library

Printed and bound by SIA PNB, Latvia

Title page image:
Grassland composition by Morna Henderson,
watercolour (contemporary).

Imprint page image:
Dog rose by Coral Prosser,
watercolour (contemporary).

Contents page images:
Scottish landscapes, Carolann Alexander,
watercolour (contemporary).

Acknowledgements page image:
Coltsfoot by Eleanor Christopher,
graphite and watercolour on vellum (contemporary).

Front cover images:
Left: Ivy by Jessica Langford,
watercolour (contemporary)

Centre: Fly agaric by Nichola McCourty,
watercolour and graphite (contemporary)

Top right: Sloe by Margaret Walty,
acrylics (contemporary)

Bottom right: Hawthorn by Margaret Walty,
acrylics (contemporary)

Back cover image:
Thistle by Clare McGhee,
watercolour (contemporary)

Contents

Foreword 7

Introduction 8

Chapter 1:
Seashores 23

Chapter 2:
Wetlands 39

Chapter 3:
Grasslands 67

Chapter 4:
Woodlands 89

Chapter 5:
Moorland and Mountains 133

Chapter 6:
Human Habitats 155

Glossary 175

Bibliography 177

Index to Scientific Names 178

Index to English Names 180

Index to Gaelic Names 182

Acknowledgements 183

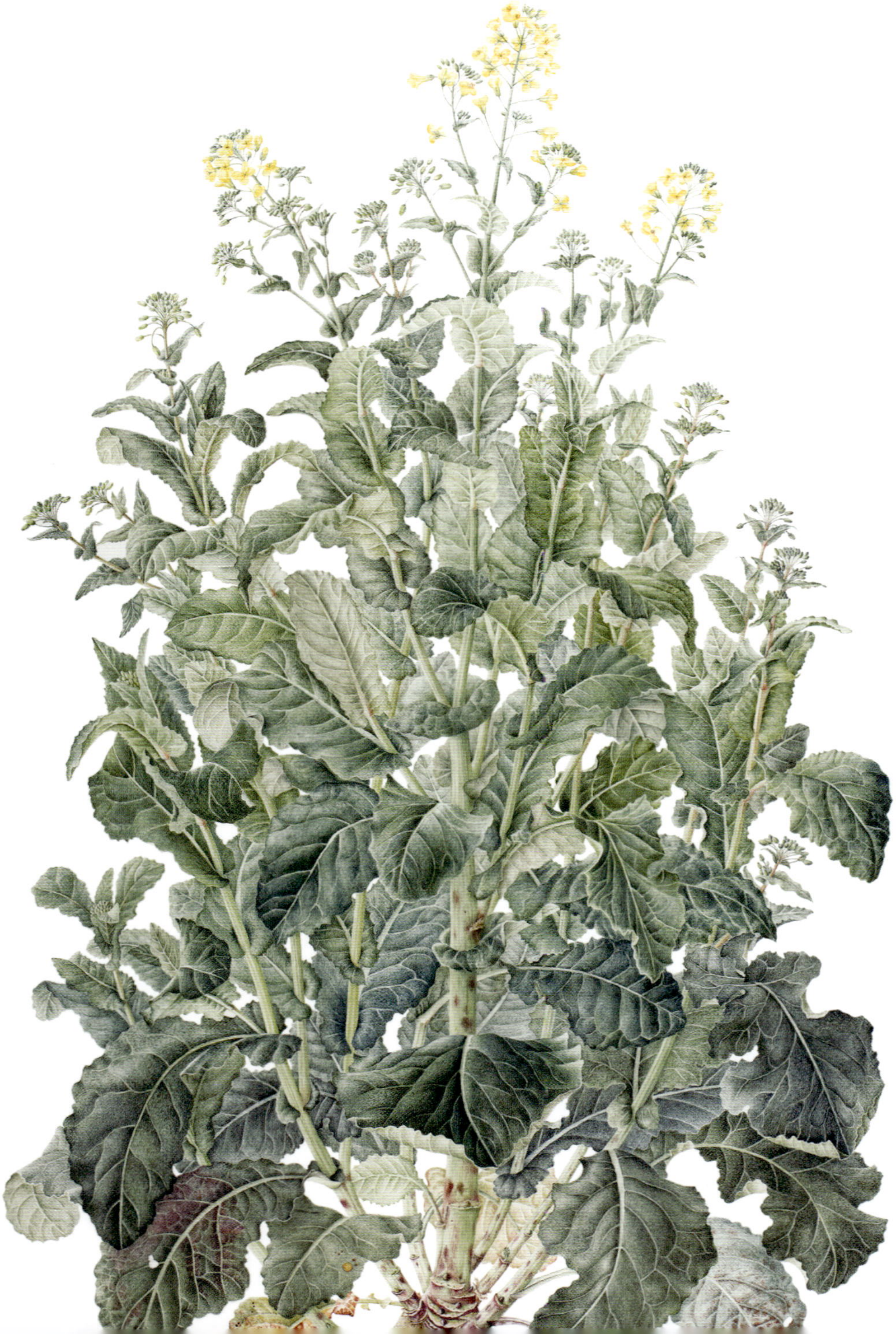

Foreword

In this book, we celebrate the plants that have shaped human lives in the place we now call Scotland; from the crops that we eat and the ornamentals that we grow in our gardens, to the weeds on urban streets and the profusion of species that define our wild environments.

We do this primarily through one of the most engaging and inspiring mediums available to us: botanical illustration.

A botanical illustration is an idealised yet scientifically accurate representation of an individual plant that includes all of the distinguishing features of the species in a way that is not always possible to capture in a photograph. The Royal Botanic Garden Edinburgh (RBGE) is a centre of excellence for this art form, serving as both a workplace and a place of learning for highly skilled artists from around the world.

The idea for this book originated from the *Botanical Art Worldwide* exhibition in 2018. Initiated by the American Society of Botanical Artists and involving the Scottish Botanical Art Collective, RBGE and more than 20 other international partners, this was a collaborative project that invited artists to capture the beauty of plants from their native flora in order to create an online exhibition that brought together the best in botanical illustration from across the globe while also highlighting plant diversity.

In addition to many of the original illustrations from the Scottish *Botanical Art Worldwide* exhibition the book brings together images from the historical collections in the RBGE Library and Archives and specimens from the RBGE Herbarium. Together with our world-renowned Living Collections that flourish in our Gardens at Benmore, Dawyck, Edinburgh and Logan, the Library and Herbarium supports the work of RBGE scientists and horticulturists in understanding and conserving plant diversity in Scotland and beyond.

All of the contributing artists and Gregory J. Kenicer are congratulated for providing such an inspirational and stunning book. We spend much of our time promoting the need to research and conserve plants around the world to protect our natural capital, and it is good to be reminded of the magnificence of our own local flora.

Simon Milne, Regius Keeper, RBGE

Left: Sutherland kale by Lizzie Sanders, watercolour (contemporary).

> **Global Strategy for Plant Conservation**
> The creation of this book is part of the Royal Botanic Garden Edinburgh's commitment to the Global Strategy for Plant Conservation. This is an ambitious, sixteen-target international plan to support habitats, plants and our understanding of their potential. The aim of Target 13 is to halt the decline of plant resources and indigenous and local knowledge about plant use.

Introduction

Human life depends on the Green Kingdom. At the most basic level, we would be unable to survive without the carbohydrate and oxygen that plants provide through photosynthesis. As building material, fuel, food and medicine, plants were an essential resource for our earliest hunter-gatherer ancestors. Even in the modern age, and despite great advances in technology, almost all human communities continue to rely on plants for their existence. Plants also enrich our lives. Whether growing in habitats created by human activity or in their natural habitats, they never cease to delight and surprise us with their beauty and diversity.

As an illustrated flora, *Scottish Plant Lore* is a modern counterpart to the herbals of the Middle Ages and Renaissance and the botanical books of the Victorian era, in that each plate is accompanied by a description of the plant's uses and associated folklore. Historical manuscripts such as these provide abundant evidence for a rich tradition of plant uses and plant-related folklore in Scotland. Historical texts are especially valuable, because they provide the most accurate information about how people understood and used the plants discussed in this book. This is supplemented by information gleaned from Scotland's rich oral history and deduced from the range of artefacts gathered across the country.

Left and right: Hazel by Jenny Haslimeier, watercolour (contemporary).

Using this book

Scottish Plant Lore is not intended as a comprehensive reference work that lists all useful Scottish flora plant by plant. In a book this size, we cannot hope to include all the fascinating information about plant use in Scotland. Instead, we have focused on species whose stories perhaps best represent interactions between plants and people in this country. These are arranged by habitat to frame their place in the Scottish environment.

This arrangement does, however, present a problem, because many plant species thrive across more than one type of habitat. In cases of difficulty trying to assign a species to a single habitat, we have been generally led by the work of the artists. For example, readers seeking information on dock (the large *Rumex* species) will find an illustration of the very rare loch-side species *Rumex aquaticus* in the 'Wetlands' chapter. However, the accompanying text mostly describes uses for the more common species, such as *Rumex obtusifolius*, which is found in habitats shaped by human activity.

We hope that this book will help encourage an appreciation of botanical illustration as well as an interest in plants and their traditional uses.

For readers who wish to learn more about plants and people in Scotland, we have provided a bibliography at the end of this chapter.

Disclaimer

Plant life supports human life. However, plants can also harm and kill. A significant minority are toxic and, if eaten, produce symptoms ranging from mild nausea to death. Certain plants irritate the skin when handled.

This book is full of descriptions of the weird and wonderful traditional uses of Scottish plants. Some of the plants were used as poisons. Others served as folk medicines, although they were probably far more harmful than curative. Our ancestors' use of plants was based on knowledge and practice accrued and developed over generations. This, in turn, was founded on their understanding and interpretation of the world around them, which differ in many ways from our beliefs today. The herbal remedies used by our ancestors were not subject to the rigorous testing of efficacy and safety that modern medicines must undergo. There is a reason that people in the Middle Ages lived half as long as we do today! Furthermore, several of the traditional uses described are illegal or require illegal actions, for example to obtain the ingredients for a love potion is a case in point (see under Foxglove in the 'Woodlands' chapter).

This book is **not** a manual on how to use plants, and it must not be used as such. In the interests of safety, we do **not** endorse the use of plants in the ways described herein. We cannot assume any liability for injury to persons or damage to property arising from such use.

Wild collection: the legalities and the alternative

Foraging for wild foods has undergone a huge increase in popularity over the past decade. However, it should be done only by people who are confident in their ability to correctly identify edible plant species. Misidentification or misuse of plants has the potential to cause illness, and in extreme cases, death.

Legally, provided the landowner has granted permission, most plant species can be collected from the wild.

The exceptions are the particularly rare species that are protected by law (Schedule 8 of the Wildlife and Countryside Act, 1981) and must not be harvested. However, regardless of whether or not the plant you wish to collect is in this category, never collect more than about 5 per cent of the individual plants or their seed in the area. By adhering to this rule, you will help ensure that wild populations of plants remain healthy and viable.

To avoid the negative effects of excessive wild collection on plant communities, try cultivating the plants you wish to use in a garden or an allotment. Many native Scottish species are commercially available from specialist seed companies and are no more challenging to grow than any other garden plants. Growing your own allows you to obtain enough of the plant to meet your needs. You will also be helping to support the wildlife it attracts.

Botanical illustration

Plants have inspired artists since ancient times. Across the world, people have depicted plants in all manner of visual media. Often, the plants themselves have been part of the medium as pigments or surfaces. In many cases, the material has degraded, and we have doubtless lost many treasures over the ages.

Botanical illustrations in medieval herbals, and in the scientific work of the Enlightenment and Victorian eras in particular, were essential for plant identification. They conveyed the key characteristics of each species, recording fine details in a way that could be easily reproduced and universally understood. Botanical illustrations continue to be used as an aid to identification, for example in the field of biodiversity science for descriptions of new species.

Recent years have seen an explosion of interest in the practice of botanical illustration, as an increasing number of people discover the joy to be derived from drawing and painting plants. Many artists are drawn to the combination of science and aesthetics. Botanical illustration is a way to connect with nature. It can also be a meditative process, demanding close observation and examination of plant material and experimentation with different techniques.

The work of many talented artists can be viewed at exhibitions across Scotland. In this country, botanical art is promoted by groups such as the Scottish Society of Botanical Artists, whose members share their knowledge and expertise through painting days, workshops and exhibitions. The Edinburgh Society of Botanical Artists is the alumni association for graduates of the Royal Botanic Garden Edinburgh's diploma course in botanical illustration.

Native and non-native plants: the botanical background

Until around 12,000 years ago, Scotland was in the grip of the most recent ice age. As the glaciers receded, tundra vegetation took hold, consisting mainly of low shrubs such as dwarf birch (*Betula nana*), scrub plants in the heather family, and several species of arctic-alpine willows that are now restricted to high mountains. Over the subsequent millennia, the land was colonised by further plants, their seeds or spores carried long distances by wind, water or animals, or more slowly and by small increments.

Changes in climate and in communities of plants established the diverse range of habitats we see today. Woodlands and forests covered about 80 per cent of the land by around 6,000 years ago. They came to be dominated by Scots pine in the north and west, with oak and hazel woodland through the west and more mixed woodland in the south and east. Alder and willows were the primary trees along riversides, and many specialist plants flourished in the permanently wet bogs and peatlands. Grasslands and heaths were rarer than they are today. The variety of topography and rock types in these areas provided specialised habitats in which many plants found their niche.

Species that colonised the landscape naturally are described as *native* or *indigenous*. In contrast, *non-native* or *introduced* species are those that have been brought into a region as a result of human activity. This may have been done intentionally, for example when plants were brought in for use as crops or garden ornamentals, or unintentionally, as in the case of weeds.

Scotland's traditional uses of plants relate to native species and arose within the communities of ancient Scottish peoples. However, they also reflect strong influences from the wider British Isles and northern and Mediterranean Europe. New species, and ideas on how to use them, were brought to Scotland by settlers or visitors and adopted. Many non-native introductions have thus found a home in our useful flora.

Left: Dwarf birch from Sowerby's *English Botany*, Vol. 33 (1812). This species is still found in tundra-like upland areas.

Human activities, including the introduction of some non-native species, have disrupted the natural equilibrium of native habitats in Scotland to the extent that very little truly original wild land remains. Efforts to conserve and understand these habitats are essential, especially in the context of global environmental change. However, changes to Scotland's flora and habitats also offer interesting new possibilities for individual plant species and plant communities. For example, non-native species in the countryside and urban areas, cut off from their wild ancestors, may evolve to meet the challenges of their new habitats, thereby creating new gene pools and ultimately perhaps new species.

Teams of horticulturists and scientists based at the Royal Botanic Garden Edinburgh and other research organisations are growing and studying native and non-native species found in Scottish habitats, with the aim of investigating their place within plant communities and the effects of human-induced environmental change on biodiversity and evolution. Only by understanding the intricate relationships between plants, people and habitats can we conserve the environment and work towards the sustainable use of natural resources.

Cultural context and notes on sources

The Mesolithic: hunting and gathering

Scotland's first-known settlers were a nomadic or seminomadic Mesolithic (literally, Middle Stone Age) society. They were able to make sophisticated stone tools, which they used to hew boats from a single log (often oak). These dugouts were probably used to travel around the coasts and up rivers in search of resources; as hunter-gatherers, Mesolithic peoples relied on what food they could obtain from foraging. Archaeological finds suggest that game and shellfish such as mussels and limpets were key components of their diet. Hazelnuts were another important food source, and deposits of the charred remains of these nuts have been found at coastal and island sites.

Discoveries such as these highlight a challenge in our interpretation of archaeological finds. The hard parts of nuts and shellfish preserve well, whereas non-woody plant material degrades easily. Because there are no written or oral records from the time, we are over-reliant on such archaeological finds, and this produces a skewed and incomplete picture of what our earliest ancestors ate and used. Modern societies with a similar way of life provide some insights into the lives of Mesolithic hunter-gatherers in Scotland, but the truth is that much is uncertain. It is generally assumed that they used almost any plant they could in versatile and inventive ways, but without hard evidence this is largely speculation.

In the Mesolithic, the coastal areas of Scotland offered many edible and otherwise useful plants, from seaweeds to flowering plants. Much of the inland area of Scotland was covered by the Great Caledonian Forest, a vast expanse of mixed woodland in which most of our native species of trees were present by 4,000 BCE. Travel through this wild wood filled with bears, boars and wolves would have been difficult. However, the people of the time were able to use the forest as a source of useful plants, building materials and fuel as well as a hunting ground for game.

The Neolithic: the first farmers

The start of the Neolithic (or New Stone Age) is defined by the introduction of agriculture. It seems to have taken about 5,000 years for the practice of agriculture to travel to Scotland from the nearest centre of origin, the Fertile Crescent in the Middle East. The reasons for this delay are difficult to determine. Specialised land races of crops suited to the climate of Scotland might have needed time to develop and become established. Alternatively, the hunter-gatherer way of life may have been simply too easy for the small populations living in Scotland at the time, and sufficient for their needs.

Once agriculture had taken hold in Scotland, by around 4,000 BCE, a wide range of introduced crops were being grown, including emmer and bread wheats and barley. By providing potentially more reliable sources of food, agriculture allowed longer-term settlement and the development of more structured, hierarchical societies in which rank and status came to the fore. These changes were accompanied by advances in the manufacture of ceramics, the crafting of jewellery, and the production of textiles woven from plant fibres and dyed with plant-derived pigments. There is no reason to assume that the shift towards an agricultural society meant an end to gathering. The collection of wild plants might well have remained just as important for the supply of food, medicine and other essentials.

Much of our understanding of how people lived in the Neolithic relies on archaeological finds, alongside our knowledge of similar societies living in more recent times in Europe and further afield. However, as with the Mesolithic, much is based on fragmentary remains and speculation.

The metal ages to the Middle Ages

The 3,000 years from the Bronze Age to the Middle Ages saw the arrival in Scotland of a great many people and the plants and ideas they brought with them. In the Bronze and Iron Ages, and owing to the establishment of agriculture, Scottish social structures underwent radical changes as people formed settled communities, led more sedentary lives and took on varied roles in society. By the Middle Ages, the people of Scotland were learning about the customs of those living well beyond Europe, as trade routes expanded to include the Middle East, Asia and Africa.

Left: Ground elder from Sowerby's *English Botany*, 2nd edn, Vol. 3 (1836). Ground elder (*Aegopodium podagraria*), along with sweet cicely (*Myrrhis odorata*), are two species introduced to Scotland, probably as medicines in monastic gardens.

Classical and early Irish literature give an impression of what life was like in Scotland over these millennia, providing many clues as to how both native and non-native plants were used at that time. The works of the ancient Greeks and Romans, along with many useful plants, reached Scotland via trade routes through the Mediterranean or over land. Similarly, the prose epics known as the Irish sagas crossed the Irish Sea. The arrival of Christianity in the early Middle Ages marked the appearance of the first formal herbals. The ideas in these books, and the plants mentioned in them, travelled with monastic scholars and healers and were adapted by local specialists such as the Beatons, a family of physicians living on the Isle of Mull.

Below: Bronze bust of Sir Robert Sibbald in the garden of the Royal College of Physicians of Edinburgh. Courtesy of the Royal College of Physicians of Edinburgh.

Information about people's interactions with plants in the later Middle Ages is provided by documents such as royal decrees, industrial records and even descriptions of witch trials. These, along with some fascinating archaeological finds, help us to understand how European traditions were modified to account for the climate, soils and culture of Scotland and differences in the availability of certain plants, and to identify plant uses that developed locally.

The late 1600s and Scottish Enlightenment

The Scottish Enlightenment and the years leading up to it are particularly noteworthy for students of plant use in Scotland. The achievements of Sir Robert Sibbald (1641–1722), physician, botanist and Geographer Royal, demand special mention. Moved by the famine that ravaged east coast farmlands from the Tay to Aberdeenshire from 1695 to 1699, he produced the seminal *Provision for the Poor in Time of Dearth and Scarcity* (1699). This is the original forager's guide, a study of almost all wild sources of food then available in Scotland. Curiously, however, it makes no mention of fungi of any kind.

With his friend Andrew Balfour, Sibbald established Scotland's first physic garden (the precursor to the Royal Botanic Garden Edinburgh) to supply plants for use in medicine and to train physicians in medical botany, which was the basis of their practice. He was also instrumental in the founding of the Royal College of Physicians of Edinburgh, which in 1699 published the first edition of the *Edinburgh Pharmacopoeia* (*Pharmacopoea Collegii Regii Medicorum Edimburgensium*). This lists standardised recipes for the medicines of the time. Most of the ingredients are botanical, therefore the *Pharmacopeia* is an unrivalled source of information about the use of plants by Scottish physicians in the late 17th century.

Sibbald was, in addition, a visionary anthropologist and archaeologist. He commissioned Martin Martin, a native Gaelic speaker, to travel the Highlands to document the region's geography, antiquities, customs and plant lore. Martin uncovered some delightful folklore, everyday plant uses, local medicines and beliefs. His findings, recorded in *A Description of the Western Isles of Scotland* (1703), inspired other scholars to tour the region. These included the botanists James Robertson and John Lightfoot (the original author of *Flora Scotica*, 1777), as well as the travel writers Thomas Pennant, James Boswell and Samuel Johnson.

Below: Sibbald was a key figure behind publication of the first *Edinburgh Pharmacopoeia* (1699). Courtesy of the Royal College of Physicians of Edinburgh.

The Victorian era to the mid 1900s

Information on the uses of plants in Scotland continued to be recorded throughout the subsequent Regency and Victorian eras, albeit sporadically. Throughout these times, plants were mentioned in letters, works of fiction and factual books. Statistical accounts record the natural history of Scotland, and information on plants and their uses can be found within wider discussions of the country's people, agriculture and industry. More accessible works with a narrower focus are also available, for example John Cameron's *Gaelic Names of Plants* (1893). This slim volume lists numerous Gaelic plant names, accompanied by suggestions as to their meanings and their equivalents in English. Cameron also delves into the uses of the plants themselves.

Alexander Carmichael's *Carmina Gadelica* (1900) is an extensive suite of poems, chants and sayings, interspersed with stories, all gathered from Gaelic-speaking regions of Scotland in the late 19th and early 20th centuries. It is a rich record of oral history, although it has been argued that *Carmina Gadelica* is part of the wider body of studies in the early 1900s that encompassed Celtic, Slavic and Nordic cultures, and was therefore subject to heavy romanticisation. *Carmina Gadelica* has been also been criticised for its 'colourful' translations into English and its uncertain identifications of plants. However, it remains a fascinating source of information on Scottish plant use.

In 1951, the School of Scottish Studies was established at the University of Edinburgh as a hub for Scottish ethnological and linguistic research. As part of its remit to preserve the cultural traditions of Scotland, it maintains an extensive sound archive comprising thousands of oral history recordings; these contain a wealth of plant lore that has yet to be comprehensively reviewed. The School of Scottish Studies Sound Archive, the Scottish Life

Archive (maintained by National Museums Scotland) and the Scottish Cultural Resources Access Network are treasuries containing enough material to inspire many lifetimes of research for those interested in traditional uses of plants in Scotland.

Recent literature

The decade after the School of Scottish Studies was established saw the publication of several classic works: Geoffrey Grigson's *Englishman's Flora* (1955), which list the names and uses of plants found throughout the British Isles; F. Marian McNeill's *Silver Bough* (1957), a study of Scottish folklore; and Isabel Grant's *Highland Folk Ways* (1961), which describes the customs and traditional crafts of the Scottish Highlands. Other folklore books mentioning plants appeared in later years.

Several books focusing on the plants of Scotland provide an excellent overview of their use, such as Barbara Fairweather's *Highland Plant Lore* (1984). Tess Darwin's *Scots Herbal* (1996) is a wonderfully accessible reference work. Mary Beith's *Healing Threads* (2004) contains accounts of the practice of Highland folk medicine garnered from interviews with some of its last remaining practitioners and their children. Beith discusses the use of plant medicines within the wider context of the whole tradition of Highland medical practice. Anne Barker's *Remembered Remedies* (2011) is another valuable source of interview-based knowledge, focusing on the Highlands and Islands, with some real gems of information from Orkney and Shetland. Fi Martynoga's *Handbook of Scotland's Wild Harvests* (revised edition 2015) has a more modern purview, describing practical uses of plants collected from the wild today.

The *Flora Celtica* project drew on information from many of these sources, as well as from first-hand interviews and the wider research literature, to document historical and present-day uses of plants in Scotland. It resulted in the publication of *Flora Celtica* (2004), the most comprehensive book on the topic to date and the ideal reference to have to hand for those wishing to learn about Scottish plant lore.

The names of plants

Gaelic, the various Scots dialects and English, together with the long list of local names, all contribute to the huge and elaborate vocabulary for plants in Scotland. Over the past 15 years, *Flora Celtica* researchers, along with Matt Elliot and Roger West, have compiled more than 4,000 plant names and other terms associated with plants and their uses. These offer intriguing insights into how people interpret plants and the wider natural world.

Unsurprisingly, beautiful, strange and useful plants have the most names, whereas rare, obscure or unremarkable ones have very few beyond their scientific name and a standard common English name designated by the Botanical Society of Britain and Ireland.

Local adaptations of more widely used names are common. For example, in the Gaelic-speaking world, St John's wort (*Hypericum* species) became *lus Chaluim Chille* (literally, St Columba's plant) or St Columba's wort. Historically, its antidepressant effects were interpreted as a quasi-miraculous defence against the effects on the mind of supposedly demonic influence (what we now understand as depression). Several other plants with bright yellow flowers, including bog asphodel (*Narthecium ossifragum*) and yellow pimpernel (*Lysimachia nemorum*), are also called St Columba's plants, but the reasons for this are unknown.

Variant names pop up. St John's wort's other monikers include *seud Chaluim Chille* (St Columba's jewel) and the puzzling *achlasan Chaluim Chille* (often interpreted as meaning 'the armpit package of St Columba'). Alexander Carmichael (1900) explained that the latter name derived from the practice of tucking the plant under an armpit, so that the active compounds could be readily absorbed through thin skin in this warm, sweaty environment!

Main: Herb Robert, also known as 'stinking Billy' or 'stinking Bob', by Jan Miller, watercolour (contemporary).

Right: The plant on the right of this illustration from Sibbald's *Prodromus* is oyster plant (*Mertensia maritima*) – the common name refers to the lustrous grey-green of the leaves.

The influence of religious sectarian divisions is apparent in some names. For example, 'stinking Billy' is a variant name for the strong-smelling herb otherwise known as 'herb Robert' (*Geranium robertianum*). This is thought to be an irreverent Catholic reference to the Protestant William of Orange. Interestingly, *Dianthus barbatus*, a non-native, scented ornamental, is known as 'stinking Billy' too. To confuse matters further, the Protestant faction called the *Dianthus* 'sweet William', the name we know it by today. Other plants are named after their habitat. Midden weed and midden mylies are general names for the weedy *Chenopodium* species, such as fat hen (*Chenopodium album*) and good King Henry (*Chenopodium bonus-henricus*), which often grow on rough ground, dung heaps and other 'middens'.

Primrose and hawthorn share the names 'may' and 'may-flower', referring to the month in which they flower. These names are sometimes corrupted to mey and meysie. Subtle changes in pronunciation and spelling of this sort help generate local variants on a name, thereby swelling the botanical lexicon. The common arable weed charlock (*Sinapis arvensis*) probably tops the table here, with no fewer than 15 name variants based on 'skelloch'.

Some names are simply descriptive. Stag's-horn clubmoss (*Lycopodium clavatum*) was known as fox-fit (i.e. fox foot). Although *Lycopodium* means 'wolf foot' in botanical Greek, wolves had been hunted to extinction in Scotland by the later 1700s, so people had to make do with the fox instead. Naturally, people disagreed as to which part of a fox this shaggy little moorland plant resembled, so it is also known as tod's tails (i.e. foxes' tails).

Introduction 19

Children's games and tricks are a rich source of plant names. Cleavers (*Galium aparine*) acquired the names bleedy tongues, blood-tongue and tongue bluiders from the tears to the surface of the tongue when a piece of the plant is quickly pulled out of the mouth. Other names for this species are catch-rogue, grip-grass, druppit-grass, guse grass, goosegrass, Lizzie in the hedge, loosy-tramps, Robbie-rin-in-the-hedge, stickers, sticky grass, witherspail, and the one I learned at school, sticky Willie.

A more complete review of this fascinating seam of cultural history deserves a book of its own. In the meantime, Grigson's *Englishman's Flora*, Edward Dwelly's *Illustrated Gaelic to English Dictionary* (1920 and later editions), and the plant lexicons of John Cameron (1883), Douglas Clyne (1989), Joan Clark and Ian MacDonald (1999) and Ellen Garvie (1999), are all excellent sources of information. In this book, we have generally avoided repeating too many of the orthographic variants that local dialects and misspellings have produced over the years and present the standard Gaelic name according to Clark and MacDonald (1999). We have included several Scots names, because no dedicated book on Scots plant names is yet available.

Below: The shaggy stag's-horn clubmoss among other upland plants, by Charlotte Cowan Pearson, watercolour (1868).

Further reading

Plant uses

This book can only touch on the many and varied ways in which plants have been used in Scotland, and we have been able to include only a select portion of the flora. However, many superb sources of information are available. The following are some of the main collections of plant uses and particularly useful original sources.

Barker, A. (2011). *Remembered Remedies: Scottish Traditional Plant Lore*. Edinburgh: Birlinn.

Beith, M. (2004). *Healing Threads*. Edinburgh: Birlinn. (First published in 1995.)

Bridgewater, S. & Milliken, W. (2004). *Flora Celtica*. Edinburgh: Birlinn.

Carmichael, A., edited by Moore, C. J. (1992). *Carmina Gadelica*. Edinburgh: Floris Books. (First published in 1900.)

Darwin, T. (1996). *The Scots Herbal: the Plant Lore of Scotland*. Edinburgh: Mercat Press.

Grierson, S. (1986). *The Colour Cauldron.* (Published privately.)

Grigson, G. (1955). *The Englishman's Flora.* London: Phoenix House. (Later editions also available.)

Henderson, D. M. & Dickson, J. H. (1994). *A Naturalist in the Highlands: James Robertson, His Life and Travels in Scotland, 1767–1771.* Edinburgh: Scottish Academic Press.

Lightfoot, J. (1777). *Flora Scotica.* London: B. White. (Available as facsimiles and online.)

Martin, M. (2014). *A Description of the Western Isles of Scotland Circa 1695, and a Voyage to St Kilda.* Birlinn: Edinburgh. (Originally published in 1703.)

Sibbald, R. (1699). *Provision for the Poor in Time of Dearth and Scarcity.* Edinburgh: James Watson. (Many facsimile editions available.)

There is a wealth of other sources of information for interested readers to work with, including recorded oral histories, research papers and theses. Deserving special mention are Camilla and Jim Dickson's eminently readable and careful overview of plants in Scottish archaeology, and Doreen MacIntyre's wonderful thesis on the dyestuffs of Scotland.

Dickson, C. & Dickson, J. (2000). *Plants and People in Ancient Scotland.* Stroud: Tempus.

MacIntyre, D. (1999). *The role of Scottish native plants in natural dyeing and textiles.* MSc thesis, University of Edinburgh, Institute of Ecology and Resource Management.

Names and language

Most of the specialist books on plant names relate to Gaelic names, with compilations of other common names in English and Scots dialects being scattered throughout other works. One exception is Geoffrey Grigson's *Englishman's Flora*, a superb source of information on plant names throughout the British Isles. The following books on Gaelic names often discuss the uses of the plants under the entries for the plant names. Several of the earlier works are readily available online as facsimiles.

Cameron, J. (1893). *Gaelic Names of Plants, Scottish and Irish.* Edinburgh: William Blackwood & Sons. (The second edition, published in 1900, contains considerable revisions.)

Clark, J. W. & MacDonald, I. (1999). *Ainmean Gaidhlig Lusan* [Gaelic Names of Plants]. Ballachulish. (Published privately.)

Clyne, D. (1989) *Gaelic Names for Flowers and Plants.* Furnace, Argyll: Cruisgean.

Dwelly, E. (various editions, 1920–2001). *Faclair Gaidhlig Gu Beurla Le Dealbhan* [Dwelly's Illustrated Gaelic to English Dictionary]. Glasgow: Gairm Gaelic Publications. (2001 edition.)

Garvie, E. I. (1999). *Plants, Fungi and Animals: Gaelic Names with English and Scientific Equivalents.* Skye: Clò Ostaig.

Identification guides

If you want to hone your plant identification skills, groups such as the Botanical Society of Scotland, the Botanical Society of Britain and Ireland and Plantlife organise regular field days and training courses. The Royal Botanic Garden Edinburgh and the Field Studies Council also provide training courses, and the latter produce a series of handy fold-out identification guides.

Of the many excellent guides to the identification of Scottish native plants, the following are among the best.

Averis, B. (2013). *Plants and Habitats.* Published by the author. (A very accessible account of plants and how they fit into their habitats, with excellent identification tips for a wide range of plants.)

Rose, F. & O'Reilly, C. (2006). *The Wildflower Key*, revised edition. London: Frederick Warner. (An accessible, well-illustrated guide with keys to many of the common wildflowers and trees of the British Isles.)

Stace, C. (2010). *New Flora of the British Isles*, 3rd edition. Cambridge: Cambridge University Press. (A weighty but comprehensive tome for the specialist.)

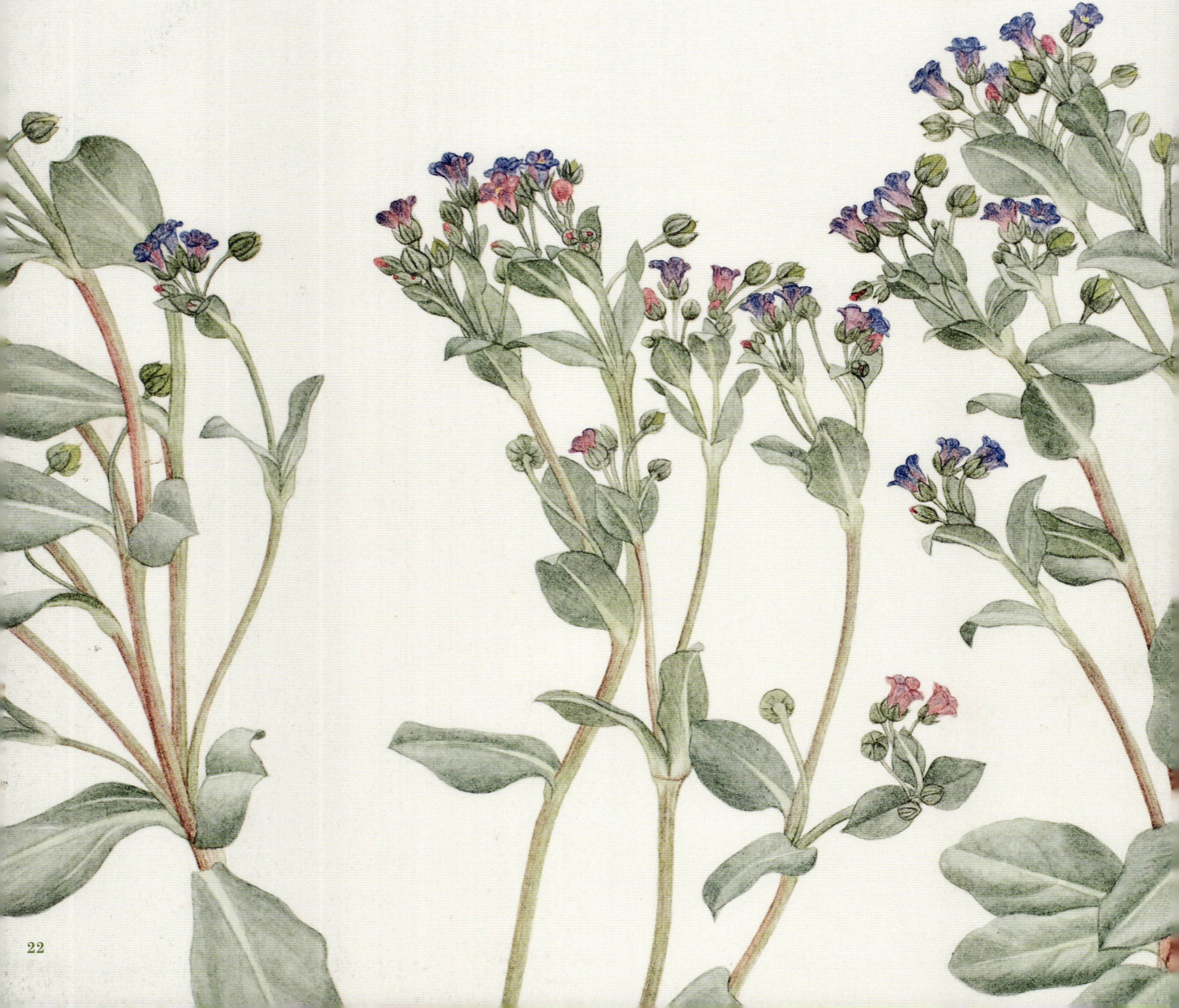

Chapter 1
Seashores

The seashores are a boundary between two realms. The salt-laden air, exposure to storms, and the sandy or rocky substrates are a challenging environment for many plants, but these areas are also often the warmest. This means that many specialised 'highly adapted' plants, including some real rarities, are found around seashores. There is also a strong influence from the areas immediately inland, which creates a diverse and particularly interesting flora.

Crucially, the seas provide transport routes for humans, and have done for millennia. Mesolithic hunter-gatherers moved around the coasts in log boats and there is good evidence of early settlements in places such as Morton Lochs, near Tentsmuir in Fife. The coasts were an abundant larder for seafood, including seaweeds, and plants from the nearby shores. The coastal regions continued to be a major source of edible plants throughout Scotland's history, and also provided medicinal and otherwise useful plants. In addition, many non-native plants were brought to coastal areas and became established due to the relatively warm and sunny climate.

Echium marinum, the plant now known as *Mertensia maritima* or oyster plant, is mentioned in Robert Sibbald's 1684 *Scotia Illustrata sive Prodromus Historia Naturalis*, the first natural history of Scotland. The plant has been little-used, except as an occasional salad leaf and specialist ornamental plant. It is now protected from collection by law, and is a subject of RBGE's extensive research and conservation projects on native Scottish plants.

Left: Oyster plant, a relative rarity from shingle beaches around Scotland's coasts, by Kathy Pickles, watercolour (contemporary).

Marram grass
Ammophila arenaria
Poaceae/Gramineae (grass family)

English:	Bent grass (Shetland), Matweed, Sea marram
Gaelic:	Muran
Flowering:	Late summer
Habitat:	Foreshore dunes. The tough, running rhizomes bind the sand together and help to stabilise the dunes.

Marram grass is a very distinctive perennial grass found on sand dunes. The slender, rolled-up, bright-green leaves help prevent it losing valuable water to the dry, salty coastal air. This adaptation has made it well-suited to sand dunes around the world, and it has become invasive in California and New Zealand.

As with many grasses, the tough leaves contain rough silica granules, so marram was used as a pot cleaner. However, its most celebrated role is perhaps as a thatch. Thatching with marram grass is usually done in two layers. The lower layer has the tips of the leaves pointing towards the apex of the roof and the upper layer the other way round, with the leaf tips pointing down, towards the ground and channelling rainwater downwards.

In Shetland, marram was considered one of the best plants for making the cords used in basketry – both for the body of the baskets and for binding them. It was also used for making brushes.

The other common large grass in sand dunes is lyme grass (*Leymus arenarius*), often mistakenly referred to as marram, but recognisable by its broad blue-green leaves. The 18th century botanist, James Robertson, noted how important lyme grass was on the west coast for its stabilising and binding role, and locals were forbidden to collect other species from the habitat so as not to disturb it.

Left and right: Marram grass specimen, RBGE Herbarium, http://data.rbge.org.uk/herb/E00907068.

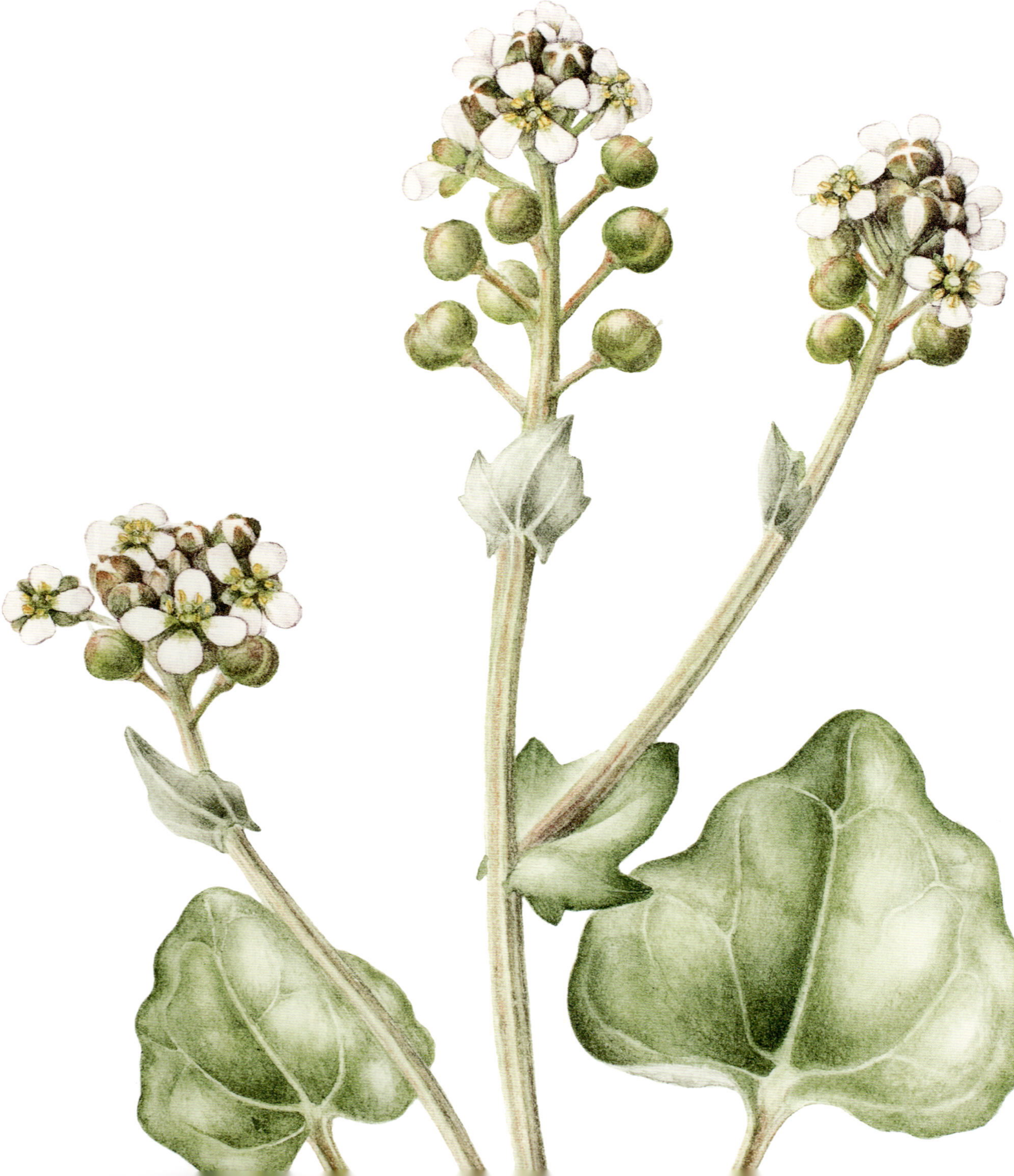

Scurvy-grass *Cochlearia* species
Brassicaceae/Cruciferae (cabbage family)

Scots:	Screeby, Screevie-grass
Gaelic:	Carran – general name for scurvy-grasses
Flowering:	May to August
Habitat:	Sandy and stony beaches, as well as path sides, roadsides and other disturbed areas where salt is found. Coastal or even inland roads that have been salted against winter ice.

This biennial herb grows to 20 cm tall, with an often messy rosette of spoon-shaped, very glossy, hairless leaves with long petioles (leaf stalks), and entire (untoothed) margins. The flowers have four white petals in the shape of a cross (hence Cruciferae), and each measures around 1 cm across. The fruits are a spherical pod, often smelling of mustard oil when crushed.

Scurvy-grass might be considered a 'superfood' if popular today. However, despite being a useful foodstuff with medicinal benefits, its extremely strong peppery flavour may be too much for the modern palate. The vitamin C in the plant staves off scurvy, and it would have been familiar to James Lind, the pioneering Edinburgh physician who investigated antiscorbutics (drugs that prevent or cure scurvy). Indeed, this species was a probable inspiration for early experiments growing brassicas on blankets on naval ships, as a source of antiscorbutics. Ultimately, citrus fruits, which preserve better, became the standard in the British Navy.

Mary Beith (2004) explained that in Scotland scurvy-grass was cultivated to a degree, and was often eaten at breakfast. A quote from Martin Martin (1703) explained the traditional view on scurvy-grass.

This rock affords a great quantity of scurvy-grass of an extraordinary size, and very thick; the natives eat it frequently, as well boiled as raw: two of them told me that they happened to be confined there for the space of thirty hours, by a contrary wind; and being without victuals, fell to eating this scurvy-grass, and finding it of a sweet taste, far different from the land scurvy-grass, they ate a large basketful of it, which did abundantly satisfy their appetites until their return home. They told me also that it was not the least windy or any other way troublesome to them.

Beyond its role in treating scurvy, there were a wide range of medical uses for the plants as part of a poultice for boils and cramps. Lightfoot (1777) said that, "The Highlanders esteem it as a good stomachic."

Boiled as an infusion, sometimes with the seaweed dulse, it was said to treat distemper (perhaps a mucous-rich cough) and costiveness (constipation).

Left: Scurvy-grass by Sarah Roberts, watercolour (contemporary).

Wild carrot *Daucus carota*
Apiaceae/Umbelliferae (carrot family)

English:	Bird's nest
Scots:	Mirrot
Gaelic:	Curran fiadhain
Flowering:	Late summer
Habitat:	Sandy soils by coasts

This biennial herb grows to about 60 cm tall. The very feathery leaves and a cluster of feathery bracts at the base of the umbels are quite distinctive. The red central flower that often emerges from the hundreds in each flowerhead really characterises this species.

The roots of wild carrots are far smaller than the hugely distorted orange vegetables we are used to today, but they are just as palatable. They have long been an important food in coastal areas, and as Lightfoot (1777) explained, "The Highlanders frequently eat the roots of the wild carrot and esteem them wholesome and nutritive."

There is a long tradition with records from the 17th to 20th centuries that women on North Uist would present the men with wild carrots following the Michelmas horse races at the end of September. Martin Martin (1703) said the gift included brightly coloured garters as well. Martin also reported that carrots were used in beer, both as a flavouring and medicinally to combat "the gravel" (small kidney stones). Thomas Pennant (1774) mentioned carrots being used in poultices to treat ulcers and cancers.

According to Johnson (1862), the leaves yield a blue dye.

Left: Field illustration of wild carrot by Jane Wisely, watercolour (1973).

Scots lovage *Ligusticum scoticum*

Apiaceae/Umbelliferae (carrot family)

English:	Scotch parsley, Sea parsley
Scots:	Luffage, Shemis
Gaelic:	Sunais
Flowering:	July
Habitat:	Rocky coastal areas, including cliffs

This perennial herb grows to around 50 cm tall. The shiny, parsley-like leaves smell like celery. The base of the leaves are inflated and often flushed with pale purple. The flowers in the umbels are very small and greenish white.

This plant is uncommon, but when found it was popular as a salad vegetable or boiled as greens. Mary Beith (2004) explains that a broth made from lamb and lovage was used to treat people suffering from an uncertain condition known as 'glacach', which is described as either a form of consumption or a swelling in the palm of the hand. She also suggests it was used as an aphrodisiac, and earlier writers ascribe a wide range of other uses to the plant.

> Shunnis is a plant highly valued by the natives, who eat it raw, and also boiled with fish, flesh, and milk. It is used as a sovereign remedy to cure the sheep of the cough. The root taken fasting expels wind. It is not known in Britain except in the north-west isles and some parts of the opposite continent. Mr James Sutherland [the first Superintendent and then Regius Keeper of RBGE] sent some of it to France some years ago.
>
> Martin Martin (1703)

> In the spring the Dairy maids give to their calves an infusion of the *Ligusticum Scoticum* and the *Rhodiola Rosea* [*Sedum rosea*, roseroot] to purge them, they call both these Lus-nan Laogh, i.e. the Calf-herb.
>
> Robertson (1869)

> The root is reckoned a good carminative [a substance consumed to expel wind]. An infusion of the leaves in whey they give to their calves to purge them.
>
> Lightfoot (1777)

Left: Field illustration of Scots lovage by Jane Wisely, watercolour (1974).

Minor species of seashores

Thrift (*Armeria maritima*) is a chirpy little plant from rocky shores, growing on rocks surrounded by yellow *Xanthoria* and other lichens. The small, bright green leaves form low, dense tussocks above which the characteristic flowerheads bob in the onshore wind. It is widespread and very common in the right habitat, but is little-used. The flower stems have been used for weaving small ornamental baskets.

Left and below: Thrift by Fran Thomas, watercolour (contemporary).

Above and right: Sea buckthorn by Sarah Roberts, watercolour (contemporary).

Sea buckthorn (*Elaeagnus rhamnoides*) is a common, but striking spiny shrub in coastal areas. The slender, simple silvery leaves are covered in tiny rust-coloured scales, but it is the bright orange berries clustered along stretches of branches that are the most distinctive. It makes an excellent windbreak and shelter belt, so most stretches appear to have been planted. There is some debate as to whether or not it is native to Scotland, and Lightfoot's otherwise comprehensive *Flora Scotica* (1777) does not appear to mention it. The tart berries are very rich in vitamin C, and have become popular in recent years as a wild food, particularly for ice creams and sorbets.

The plant forms a symbiotic association with a fungus from the genus *Frankia*, which inhabits specialised nodules in the roots. The *Frankia* fixes nitrogen from the air and provides it to the plant, thereby conferring sea buckthorn with a major advantage on the nutrient-poor sandy soils in which it grows. *Elaeagnus rhamnoides* is host to an exclusive bracket fungus called *Phellinus hippophaeicola*.

Chapter 1: Seashores 31

Seaweeds

Seaweeds are among the most versatile coastal plants, and historically provided foods, medicines, fertilisers, magic ingredients and tools, and had roles in both economic miracles and economic disasters. Although we often call them plants for convenience, they are not, by strict definition, true plants. Seaweeds are a mixed bag of more 'primitive' organisms; they do not have any vascular system or roots, and most of their tissues are simple, allowing them to absorb minerals and other nutrients directly from the surrounding seawater.

Seaweeds are often recorded as being 'famine foods', but many species were eaten whenever available, particularly as an accompaniment to shellfish and fish. Robert Sibbald's wonderful guide to wild foods in Scotland, *Provision for the Poor in Time of Dearth and Scarcity* (1709), records the following uses of seaweeds as 'famine food'.

> In the North, and in many other places of the coast of this Countrey, People feed upon Slake, that is, the sea lettuce; They make Broath with it and sometimes serve it up with Butter: it groweth upon the Rocks washen with the Sea. Some of them eat much Dills; the Fucus Membranaceus Ceranoides C.B. and some eat that sort of Sea Tangle,

Left: Seaweeds by Margaret Walty, acrylics (contemporary).
Right: Bladderwrack by Anne Dana, watercolour (contemporary).

called Fucus Nostras latissimus, Tenui Folio. It is of a pleasant taste betwixt salt and sweet: it's Eaten as a salad.

This was written before the introduction of a formal system of botanical nomenclature, therefore it is not fully clear which seaweeds Sibbald is describing. However, the passage is good evidence that sloke (also known as laver, *Porphyra umbilicalis*), and perhaps the sea lettuce (*Ulva lactuca*), dulse (*Palmaria palmata*) and sugar-wrack (*Saccharina latissima*), were all being used as food in the 17th and 18th centuries.

Vernacular names of seaweeds can be confusing. Tangle is usually the name for the large *Laminaria* seaweed, but can be applied much more widely to the large brown seaweeds as a whole, including *Saccharina*, *Fucus* and *Ascophyllum* species. Similarly, ware is a general name for seaweeds, although it is usually applied to the larger brown species. Wracks are the large brown species of the genus *Fucus*, such as bladder wrack and spiral wrack. Kelps are the largest brown species from the genera *Laminaria*, and *Saccharina*, often from deep waters.

Kelp is also, however, the name for the product of burning many different species of brown seaweed to produce a slag-like, mineral-rich ash. Throughout the later 17th and 18th centuries and into the earlier 19th century, seaweeds were burned in pits on the foreshore as part of the kelp industry. The ash is alkaline, and was a key source of soda and potash. It was used for bleaching linen, for making soda glass and soap, and as a fertiliser, among other things. The kelp was very profitable for the landowners, but the daily toil of keeping the kelp-pits burning brought little return to the crofters, who were diverted from their work tending the land.

Perhaps inevitably, the bottom fell out of the industry, first when industrial-scale guano collection from South America allowed guano to displace kelp ash as a fertiliser, then when a chemical process for producing iodine (another kelp-ash derivative, used medicinally) was developed. The former was undoubtedly a catalyst for bringing sheep onto the land, which precipitated the Highland Clearances.

The large brown seaweeds such as the wracks and kelps were used widely as fertilisers. They were added to the soil directly in the autumn or some time before the growing season, to allow the seaweed time to rot and the nutrients to be incorporated into the soil.

Carrageen/Irish moss (*Chondrus crispus*) is another seaweed species of note. This was often boiled in milk to thicken it into a junket (sweetened and flavoured curds) or set it into a pudding. Carrageenan, the algal polysaccharide responsible, is a major commercial thickening agent. Algal polysaccharides are fairly

Chapter 1: Seashores

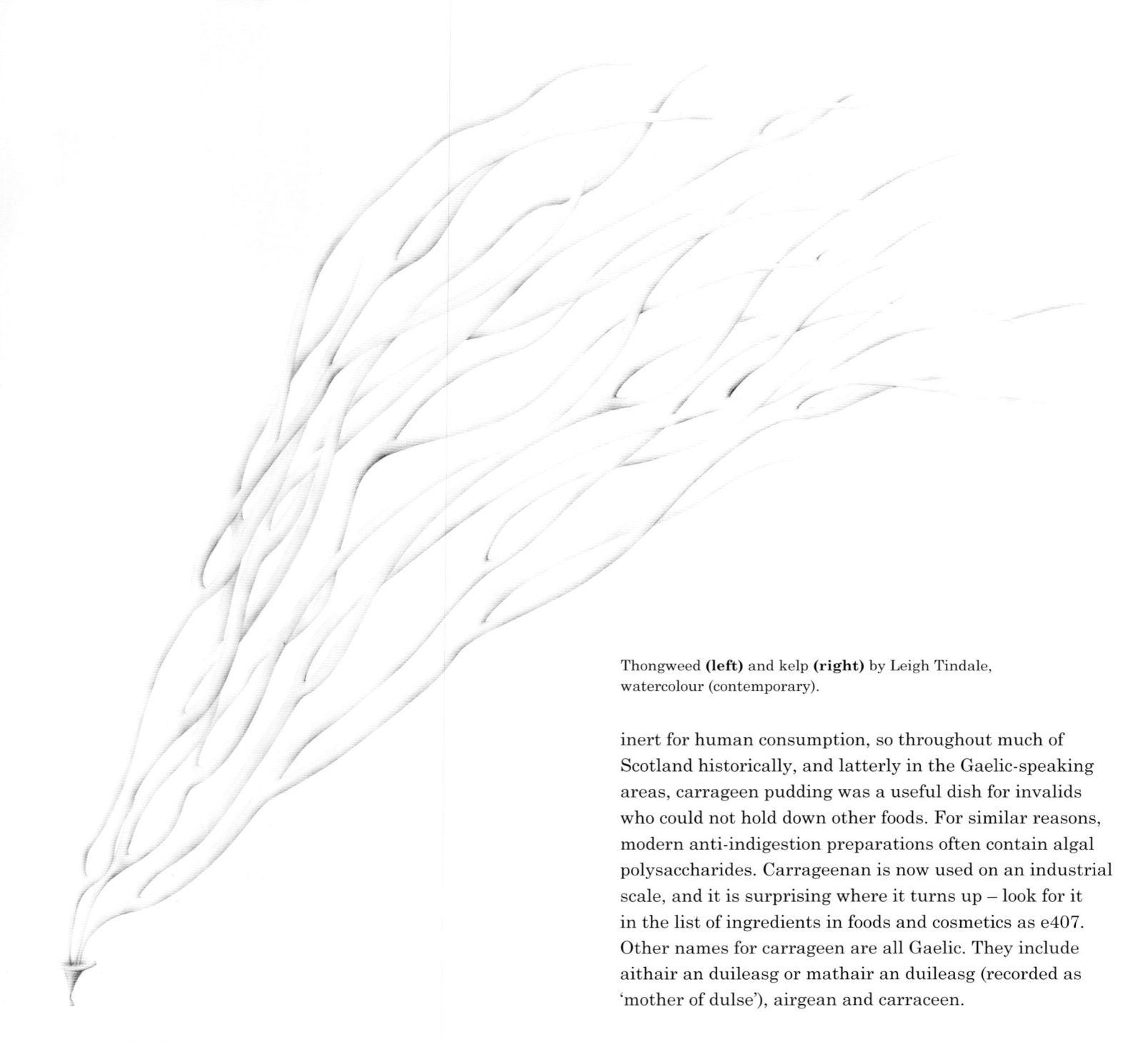

Thongweed **(left)** and kelp **(right)** by Leigh Tindale, watercolour (contemporary).

inert for human consumption, so throughout much of Scotland historically, and latterly in the Gaelic-speaking areas, carrageen pudding was a useful dish for invalids who could not hold down other foods. For similar reasons, modern anti-indigestion preparations often contain algal polysaccharides. Carrageenan is now used on an industrial scale, and it is surprising where it turns up – look for it in the list of ingredients in foods and cosmetics as e407. Other names for carrageen are all Gaelic. They include aithair an duileasg or mathair an duileasg (recorded as 'mother of dulse'), airgean and carraceen.

Several other seaweeds were used as foods, including thongweed (*Himanthalia elongata*). The perennial 'buttons' at the base of the plant were used as a thickener in lamb stews, while the reproductive structures (known as 'mirkles') on *Alaria esculenta* were eaten raw or in stews.

The beautifully named mermaid's tresses (*Chorda filum*) is unmistakable – long, slender brown locks of seaweed below the low-tide mark. The locks are up to 3 m in length and coated in a thick mucilage (gelatinous substance). Like many other seaweeds, it can be eaten – and almost certainly was. Patrick Neill (1806) recommends it as the best seaweed for yielding kelp ash. Lightfoot's Flora Scotica (1777) suggests it can be used as a substitute for "Indian Grass", perhaps meaning it can be woven into nets. This use is reflected in many of its common names: in Gaelic, it is driamlach, langadair or gille mu lean (young man's net, or rope), recalling the English "dead man's rope", and other names in English for this striking species include sea lace and cat gut. In Scots it is, of course, deid men's ropes as well as luckie's lines and lucky minny's lines.

Several modern companies are helping revive the popularity of these sea vegetables as gourmet ingredients.

Above: An exquisite red seaweed specimen, closely related to dulse, from the RBGE Herbarium.

Dulse *Palmaria palmata*
Palmariaceae (Dulse family)

English:	Hand fucus
Scots:	Reid ware, Rid ware
Gaelic:	Duileasg

Dulse is a simple red seaweed that grows to about 30 cm. It has a flat, translucent red blade that branches forwards or sideways many times from a small holdfast. It grows fairly low on the shore, on rocks, or particularly on the stems of larger submerged brown seaweeds. This species is perhaps the star performer among seaweeds in Scotland for versatility.

Martin Martin (1703) gives an intriguing account of dulse being used in obstetrics at the end of the 17th century:

> A large handful of the sea-plant dulse, growing upon stone, being applied outwardly, as is mentioned above takes away the afterbirth with great ease and safety; this remedy is to be repeated until it produce the desired effect, although some hours may be intermitted: the fresher the dulse is, the operation is the stronger: for if it is above two or three days old, little is to be expected from it in this case. This plant seldom or never fails of success, though the patient had been delivered of several days before; and of this I have lately seen an extraordinary instance at Edinburgh in Scotland, when the patient was given over as dead.

This practice may be an example of the Doctrine of Signatures, which holds that the treatment looks like the condition it is intended to cure. Another theory is that the shock of a lump of fresh, cold seaweed dropped on the stomach may have caused spasm enough to expel anything!

Alexander Carmichael's *Carmina Gadelica* records an even more quasi-magical cure for an obscure condition called 'the falling of the uvula':

> Margaret MacKenzie, "am Poll Glas," Polglass, Achiltibuie, Ross, performed the cure as follows. She went to the strand and brought thence what were called na ciochagan traghad, uvulas from the strand (dulse). She hung them above the fire, which was on the floor (by some others they were placed on the slabhraidh, pot-chain). She recited a rann (rune), and raised the fallen uvula.

It seems likely that the 'uvula of the strand' refers to a seaweed, and dulse does look rather like a human uvula.

More prosaically, dulse soup (càl duilisg) was a treatment for constipation, stomach ailments and skin complaints. Dried dulse was eaten after fasting to get rid of worms, and as with sea lettuce (*Ulva lactuca*), it was also used to ward off scurvy. Martin Martin explains how a Dr Pitcairn told him that a "cure had been performed in the shire of Fife for [a 'twisting of the guts']. A cataplasm of hot dulse, with its juice, applied several times to the belly, cured the iliac passion".

On Skye, dulse was boiled with a little butter then used as a compress to induce sweating in patients.

The Reverend Landsborough, writing in his *Popular History of British Seaweeds* (1849), records that dulse "was thought very efficacious as a sweetener of the blood, and in warding off, or curing, scorbutic (scurvy) and glandular affections". It is known to have been used as a substitute for chewing tobacco. The plant would be washed and cut into cut into small pieces then mixed with butter for flavour and dried. Landsborough extols the virtues of dulse over true tobacco:

> How much better had it been for them that [the Highlanders] had stuck to the use of the less nauseous, less filthy, less hurtful dulse. Indeed, instead of being hurtful, it is thought wholesome and not unpleasant, especially when it is eaten fresh from the sea.

In the 1860s, Johnstone and Croall record that dulse was only eaten raw in their day, except when they were young:

> some people [gave] it a slight scorching or roasting by rolling it round a heated poker, after which it had a very peculiar flavour, which to most persons, as well as to us, was very disagreeable. By this process the red colour was changed to green.

Johnstone and Croall also mention that when dulse was hawked around the streets the small crustaceans and parasites living on its surface were not removed before eating and in fact "form a delicate part of the morsel".

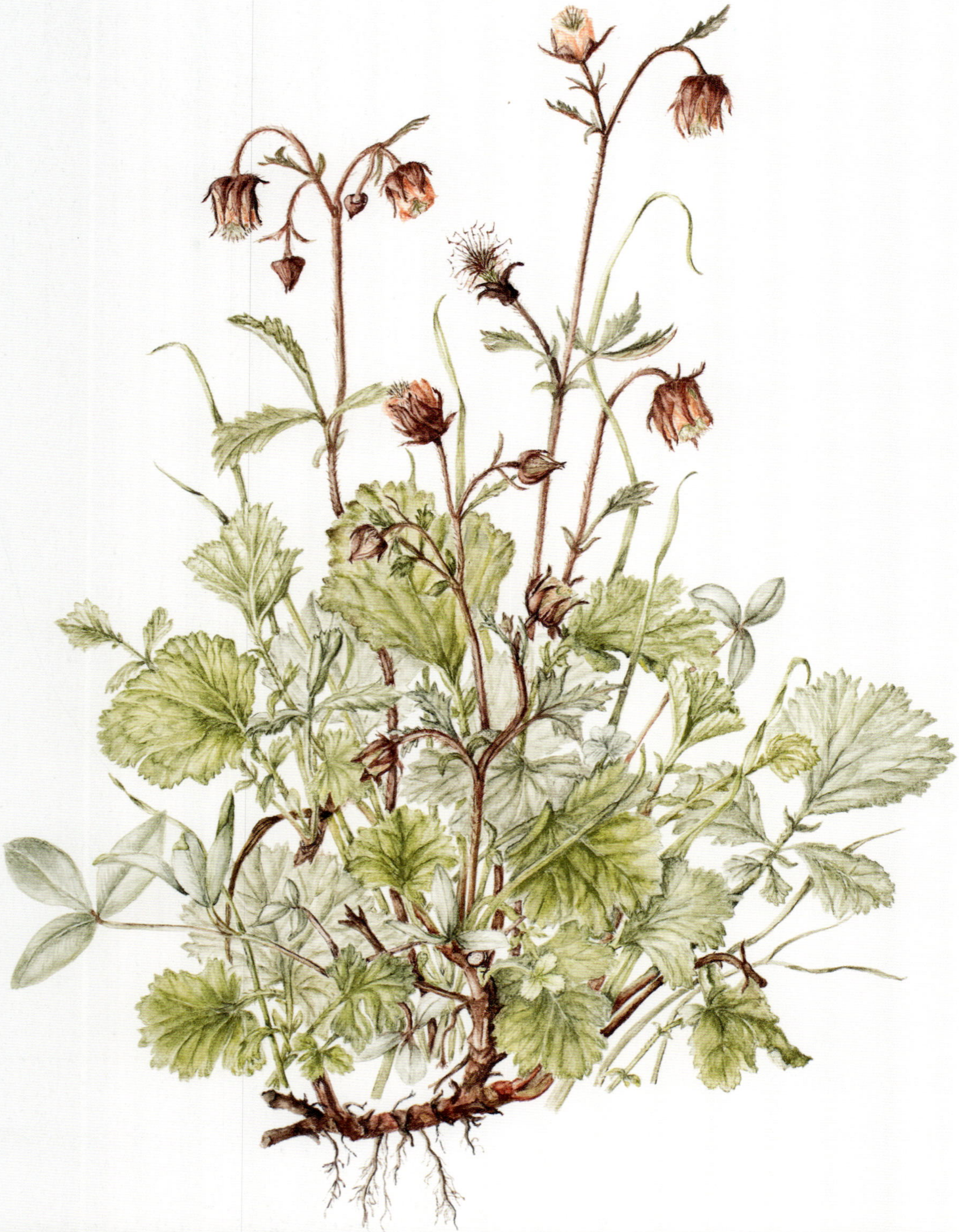

Chapter 2
Wetlands

From tarns in upland corries and the margins of deep inland lochs, through torrents and streams, bogs and mires to the large, lazy river estuaries, wetlands provide a wealth of habitats for useful plants. The rivers and lochs have long been important routes for human travel, whereas expanses of *Sphagnum*-rich bogs were seen as challenging, wild habitat, and many have been drained and built on.

Today, we have a far better appreciation of the conservation value and biodiversity of these habitats, and the immense range of ecosystem services they provide. Therefore, harvesting of wetland plants is now regulated. In the past, however, their uses ranged from the prosaic to the mystical. A common theme is their use as sources of black dyes. Tannins are common in the roots and rhizomes of aquatic and water-marginal plants, and tannic acid can be made into a reasonable (if somewhat acidic) ink or dye with the addition of 'copperas' (iron sulphate). Edmonstone (1841) explains that bog-iron was a major source of copperas, and it could conveniently be collected from boggy habitats.

Left: Water avens by Janette Dobson, watercolour (contemporary).

Left and right:
Alder by Sharon Tingey, watercolour (contemporary).

Alder *Alnus glutinosa*

Betulaceae (birch family)

English:	Scotch mahogany
Scots:	Aar, Aller, Allertree, Alrone, Arn
Gaelic:	Feàrna
Flowering:	Mainly in February and March, female fruit heads develop over the course of the year into cones reminiscent of those of a pine
Habitat:	Found by slow-flowing and still fresh water, and forming woodland in permanently wet flushes (called alder carr)

This medium-sized deciduous tree reaches around 20 m tall. The crown often forms a broad column, with a waisted profile. In winter, the tips of the branches are vibrant, giving the plant an aura of hazy purple. The leaves are glossy, inverted-heart shapes with divergent parallel main veins, and up to 15 cm across. Male and female catkins are separate, but on the same tree.

Alder is an immensely versatile tree. The wood has been used for its water resistance in a huge array of tools and construction. Bridges and walkways, piers and pontoons, including those supporting the Iron Age and medieval crannog loch-houses, were all typically made of alder. Alder water-wheels and lock gates were used in milling and transport. Clogs were almost invariably made of alder; indeed, in the south of Scotland during the early 1900s, 'Scotch mahogany' was in such demand for making clogs that birch had to be used instead.

One of the most intriguing uses of alder was to make deep tubs for storing butter, which were sunk into bogs to keep the contents cool. The National Museum of Scotland holds a beautiful example, carved from a single piece of wood that must have come from a prime tree.

Alder produces high-quality charcoal and was used in smelting and in the production of gunpowder. As with birch, it was coppiced to provide shoots for the charcoal industry.

The bark is tannin-rich, so was in high demand for tanning leather and fishing nets, and with copperas (iron sulphate) was made into a black dye somewhat like oak-gall ink. The inner bark produces a red or yellow dye, or with the addition of alum, a gold-yellow dye. In each case, the mixture should not be boiled, or the colour will be less vivid or fix poorly.

The Gaelic 'fearn' appears in many place names across Scotland, such as Fearn (Easter Ross), Glen Fearnach (Perthshire) and Caochan Feàrna (near Dalwhinnie). Ardfern and Alltfearna may be the alder-rich parts of Argyll in which Deirdre and the lovers Diarmid and Gráinne are said to have hidden from pursuers in the Irish Fenian cycle of epic tales.

The closeness and density of alder carr woodland has given the tree a sinister reputation. This is heightened by the fact that the wood reddens (as if it contains blood) when cut. This same bloody red is also attributed to God's curse on alder, as it was allegedly used for making the Crucifix. That said, many other equally innocent tree species have been similarly accused and cursed for their various strange characteristics.

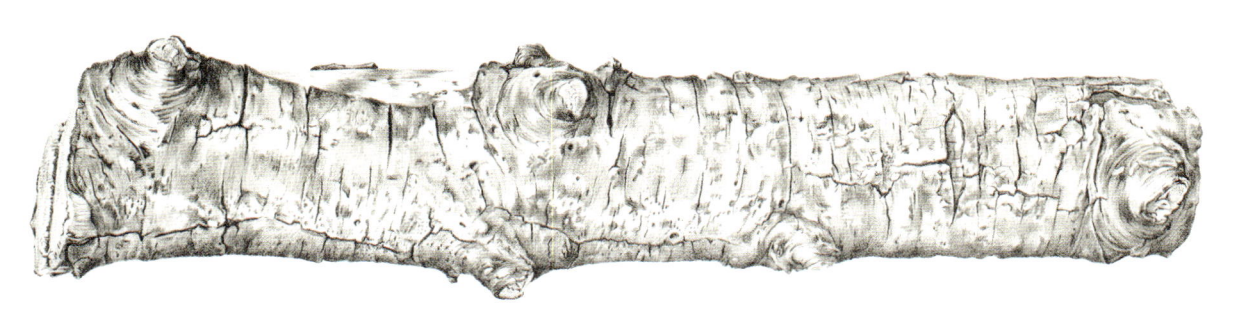

Chapter 2: Wetlands 41

Meadowsweet
Filipendula ulmaria

Rosaceae (rose family)

English:	Cu-Chullain's belt, Lady o' the meadow, Meadow queen, Meduart/Medwort (Archaic)
Scots:	Blackin-girse, Leddy o' the meadow, Yirnin' girse, Yolgirse
Gaelic:	Creas Chu Chulain
Flowering:	Mid to late summer (June to September)
Habitat:	Wet meadows and similar habitats: dune slacks, margins of lochs (both sea and freshwater), mires (sometimes forming large drifts and beautiful when flowering en masse)

This perennial herb grows to about 1 m tall. The stems and the rachis (central vein) of the compound leaves are a vivid red. The softly toothed leaflets grow between two and five pairs, about 5 cm long, with smaller leaflets between. The leaf is tipped with a ternate (three-pointed) leaflet. The small flowers form large, creamy masses atop red stems. The fruits are tiny, coiled-looking capsules. The whole plant, but especially the flowers, smells strongly of oil of wintergreen when crushed.

This striking plant has an ancient heritage of use described throughout early literature and supported by archaeological evidence.

Meadowsweet appears in brewing and wine-making traditions. It is alleged to have been an important ingredient in traditional heather beers, although it is difficult to find conclusive proof of when this practice began. Several early finds of pollen in pottery or other burial finds have been interpreted by various researchers as signs of brewing. These include a possible Neolithic beaker from Rhum and discoveries on Arran and from Bronze-Age Fife. The flowers are rich in nectar, so used as a good sugar source for brewing. Furthermore, meadowsweet's distinctive scent imparts a degree of flavour when used in modern drinks. Indeed, the common name is usually interpreted as 'mead-' rather than 'meadow-' sweet.

The smell was put to good use when meadowsweet was used as a strewing herb – scattered on the floor with other scented plants so that their pleasant smell wafted up when walked on, thereby masking the odour of unwashed bodies and livestock. This temporary carpet could then simply be swept out the door and replaced. The scent mostly arises from methylsalicilate (oil of wintergreen) – a precursor to aspirin – thus meadowsweet was used as a painkiller. Lightfoot (1777) also lists it as a treatment for dysentery.

The Gaelic name comes from its use by the Irish hero Cu-Chullain as an antidote to his 'warp-spasm', a form of berserk battle rage that required him to be bathed in a cauldron of herbs to calm him down.

As with many bog and aquatic plants that appear here, meadowsweet roots were used to produce a black dye, or ink, hence 'blackin' girse', another of its common names. Sue Grierson's *Colour Cauldron* (1986) explains how sorrel can be added to give the black a more bluish hue. Doreen MacIntyre's (1999) research project on Scottish dyes also describes the flowers being used for yellows, from lime-yellow (with copper) to pure yellow (with alum).

Left and above: Preserved specimen of meadowsweet, RBGE Herbarium, http://data.rbge.org.uk/herb/E00411449.

Chapter 2: Wetlands 43

Yellow flag iris
Iris pseudacorus
Iridaceae (Iris family)

English:	Flag, Fleur de lys, Wild iris, Yellow flag
Scots:	Cheeper, Dug's lug, Segg/Seggan (sedges), Water skegg
Gaelic:	Seileasdair
Flowering:	May to July
Habitat:	Wetlands, from freshwater- and sea-loch margins, to roadside ditches, wet sloughs and dune slacks, sometimes forming extensive beds. Clumps of iris are a common feature of the western coastal element of Scotland's flora, but the plant is widespread throughout.

This perennial herb grows to slightly over 1 m tall. The leaves arise from a thickened underground stem, the rhizome, which reaches about 5 cm in diameter and branches to form dense beds. The leaves are strap-like, long, and taper to a point, with parallel veins. The large (10 cm across) three-spoked yellow flower is very distinctive as an iris flower. This is the only iris truly native to Scotland.

When used in thatching, yellow flag is particularly suited as a layer underneath marram, because it provides an even, flat base. The leaves were also used in basketry, but only rarely, and in modern times are used mostly for ornamental and specialist display baskets.

As with other wetland plants, yellow flag was most commonly listed (and mentioned in Lightfoot, 1777) as a source of black inks and dye from Arran and some of the Western Isles. Other writers mention dark-green dyes (Johnson, 1862), blue from the rhizome (perhaps with copperas and an acid), green-brown, and pale yellow. Although this wide range of dyes has been suggested as the reason for its Gaelic name, Seileasdair (rainbow), it is more likely just a direct relation to the wider European idea of 'iris' meaning rainbow (Iris is the name of the goddess of the rainbow in Greek mythology).

Lightfoot describes a wonderful medicinal use from Mull, in which iris 'root' (almost certainly the rhizome, or underground stem) was pulverised with daisies, and a teaspoon of the resulting juice was poured into a patient's nose to treat toothache and nasal problems. Perhaps unsurprisingly, a possible side effect was catching a cold after the treatment. However, help is at hand; Mary Beith (2004) explains that a snuff made of the rhizome was used in the treatment of colds. Martin Martin (1703) mentions that "glisters" were made of salt water, butter and the "roots of flags". The meaning of glisters here is uncertain, but it is often a liquid injected into an area needing treatment, most commonly as an enema.

On a more pleasant note, 'cheeper', a Scots name for iris, comes from the squeaking whistle that can be made using the leaf as a reed between the thumbs.

Left and right: Yellow flag by Lyn Campbell, watercolour (contemporary).

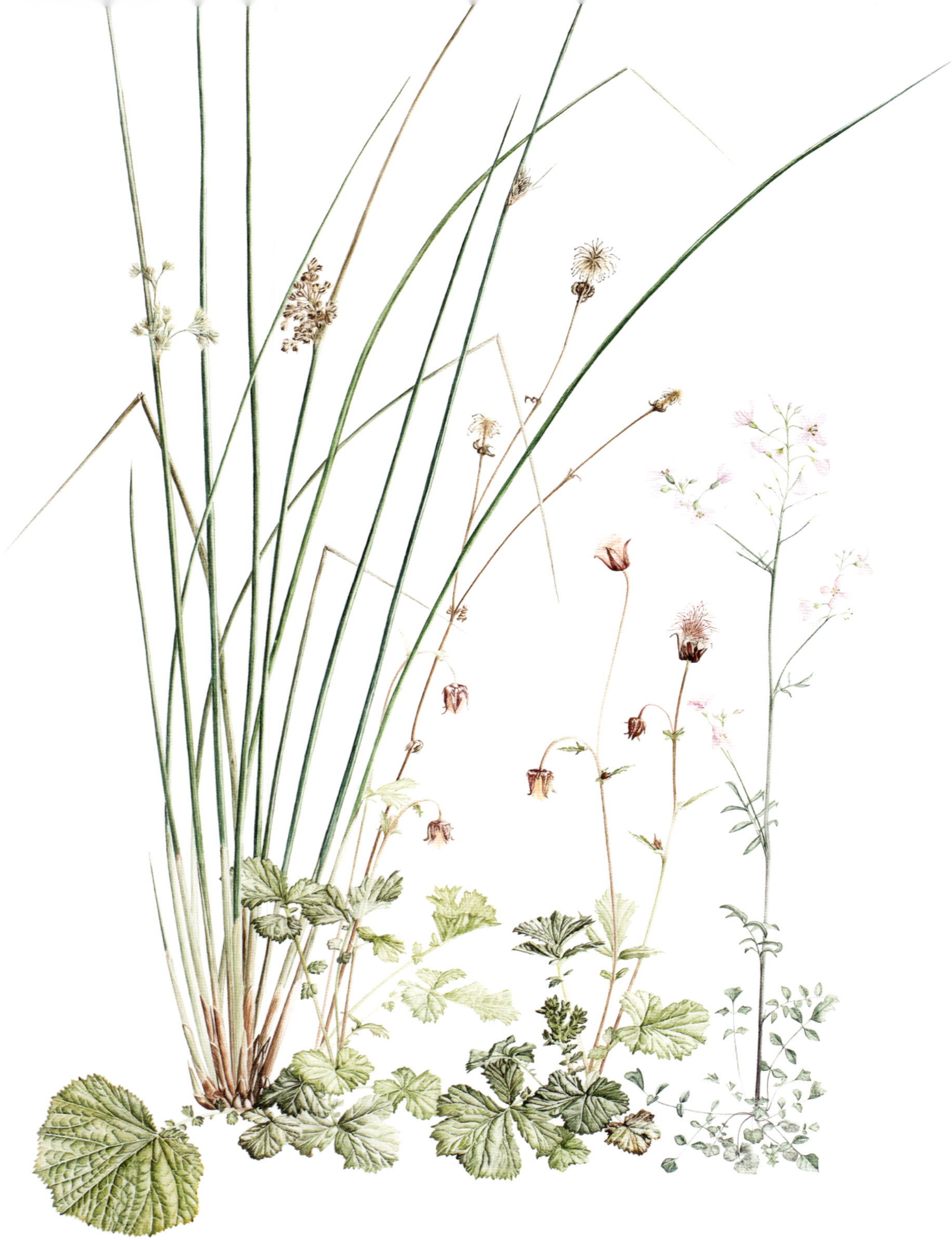

Soft rush *Juncus effusus*
Juncaceae (rush family)

English:	Rush
Scots:	Floss
Gaelic:	Luachair bhog
Flowering:	June to August
Habitat:	Wet meadows, bogs, ditches and other acid and nutrient-rich wet habitats

This perennial tussock-forming herb grows to 1 m tall. The cylindrical, dark leaves have a fairly sharp tip. Soft rush produces a spray of tiny brown flowers with chaffy brown sepals and petals. The fruits are small chestnut-brown capsules.

For centuries, common rushes and grasses have been used for making small baskets. Lightfoot records this use from 1777, and Martin Martin before him. The rushes or grasses are cut, dried and then soaked for a few hours before use. They can be woven as they are or first plaited into ropes, which are then woven into baskets. Although rushes were also used occasionally for thatching, they generally only lasted a couple of years so were seldom favoured.

Left: Soft rush with water avens and cuckoo flower by Morna Henderson, watercolour (contemporary).

Rush-wick lights were a staple source of lighting. The green rind was stripped from the outside, leaving the soft, spongy pith with a small strip of rind running up it for rigidity. This was then dipped in animal fat from either fish or mutton, and held in rush-nips (a rush light holder). When lit, the wick produced a soft flame.

The flowers produce an apricot-coloured dye with alum and cream of tartar as mordants.

Martin Martin (1703) explained an interesting medicinal use from Skye and neighbouring islands:

> To cause any part of the body to sweat, they dig a hole in an earthen floor and fill it with hazel sticks and dry rushes; above these they put a hectic stone [quartz], red hot, and pouring some water into the hole, the patient holds the part affected over it and this way procures a speedy sweat.

48 Chapter 2: Wetlands

Bogbean *Menyanthes trifoliata*
Menyanthaceae (bogbean family)

English:	Bogbean, Bog-nut, Buckbean, Marsh trefoil
Scots:	Gulsa girse, Threefold, Trefold, Water triffle
Gaelic:	Tri-bhileach
Flowering:	June and July
Habitat:	Still-water margins, and bogs in neutral and acid water

This perennial aquatic herb grows to 2 m long. The grey-green hairless trefoil leaves and flowerheads are held above the surface of still water. The flowers, which are around 3 cm across, with five white or pale-pink petals fringed on the margins, are borne in a spike of up to a dozen. The fruits resemble an upright, thick, green needle, about 5 cm long.

Bogbean was renowned as a tonic, providing general preventive and minor curative benefits as well as promoting recovery from illness and wounds. In Glencoe, it was stored in stone jars then simmered for use as a spring tonic.

Lightfoot (1777) also cites its use as a tea, particularly useful "to strengthen a weak stomach". Mary Beith (2004) lists a wide range of local uses: the leaves were used for drawing pus from wounds; on Lewis, it was used for treating tuberculosis, heart problems and asthma; and on Uist it was given for constipation. It was put to a similar use with animals too, when they had eaten too much.

'Gulsa girse', one of bogbean's common names, reflects its use in the treatment of jaundice on Shetland.

The roots (and potentially stems) yield a brown dye. The leaves can be used with heather tops, cream of tartar and alum as mordants to produce a green dye.

Left and right: Bogbean by Fran Thomas, watercolour (contemporary).

50 Chapter 2: Wetlands

Watercress
Nasturtium officinale
Brassicaceae/Cruciferae (cabbage family)

Scots:	Carse, Geraflouer, Gerofle, Kerses, Lillie, Wall-kerses, Well-girse, Wild skirret
Gaelic:	Biolair uisge
Flowering:	May to July
Habitat:	Slow-flowing streams, pools and similar bodies of water

This perennial aquatic herb grows to 50 cm tall. The leaves are compound, with heart-shaped leaflets. The flowers are small, each consisting of four white petals forming a cross.

Watercress was (and still is) a commonly harvested salad vegetable. It is mentioned in Robert Sibbald's *Provision for the Poor in Time of Dearth and Scarcity*, and was traditionally added to soups and broths, or boiled in water and strained, in the belief that it would aid recovery from colds and flu. On Mull, it was particularly valued for staving off scurvy. Although there is a risk of contracting liver flukes from wild watercress, Mary Beith (2004) wrote that the people of Uist believed this could be avoided by picking the plants from above the water line.

A violet dye can be produced from the plant.

Left and right: Preserved watercress specimen from the RBGE Herbarium.

Chapter 2: Wetlands

White water-lily
Nymphaea alba

Nymphaeaceae (water-lily family)

Scots:	Bobbins, Cambie-leaf
Gaelic:	Duilleag-bhàite bhàn
Flowering:	June and July
Habitat:	Still water – in lochans and pools ranging from neutral to acidic; often in peaty water

This perennial aquatic herb is the very distinctive white water-lily, with rhizomes in the mud giving rise to classic lily-pad leaves floating on the surface.

Despite its striking appearance, this plant is little-used. The rhizome was used to produce a chestnut-coloured dye, and with copperas it produces the classic black dye of many aquatic plants. A fragmentary record cited by Mary Beith (2004) shows that it would be boiled in vinegar and then used to treat corns.

Left: White water-lily, John Hutton Balfour teaching diagram collection (19th century).

Chapter 2: Wetlands

Above: Dissection of white water-lily flower, John Hutton Balfour teaching diagram collection (19th century).

Chapter 2: Wetlands 53

Butterwort
Pinguicula vulgaris

Lentibulariaceae
(butterwort family)

English:	Butter-plant
Scots:	Earning-grass, Ecclegrass, Ekel-girse, Ostin girse, Sheep-grass, Sheep-rot, Sheep-root, Yirnin-girse
Gaelic:	Mòthan
Flowering:	June to August
Habitat:	Nutrient-poor, acid, waterlogged bogs – usually among peat

This perennial carnivorous herb looks like a pale lime-green star. At the base there is a whorl of triangular, in-rolled, pale, yellowish-green leaves, each 3–6 cm long and with sticky upper surfaces that attract and trap insect prey, which is digested *in situ*. The solitary flowers, on 15 cm stalks and around 3 cm long, are purple and tubular, with a broad mouth and a spur behind. There is usually only one flower, but in exceptional years individual plants may produce up to a dozen.

Butterwort is a wonderful little carnivorous plant, with a diverse range of uses and lore, both good and bad.

There is an intriguing reference to the plant being "distilled for a summer drink" (*Fairweather's Highland Plant Lore*, 1984), but its main practical use was for curdling milk for butter and cheese-making – the digestive enzymes may act like rennet. As Cameron (1883) explains, "It is believed it gives consistence to milk by straining it through the leaves."

It appears to have been used on both domestic and small-commercial scales throughout Scotland. This species was ascribed supernatural powers of milk protection, perhaps an extension of its practical role. It was one of several plants that were made into little hoops and placed under buckets, jugs, pails and churns to protect the milk inside from baleful magic. Beyond this, Mary Beith explains, in *Healing Threads* (2004), that a golden amulet, containing "the dust" of this plant was thought to be very effective in combating all manner of dangers, including hunger, a broken heart, wounds received in combat, loneliness, drowning and the wiles of witches and other supernatural evils.

Alexander Carmichael (1900) also lists many protective magical properties of the plant under the name of 'mòthan', although in his characteristically 'vibrant' style he expresses some uncertainty about the accuracy of his identification.

The mòthan is one of the most prized plants in the occult science of the people. It is used in promoting and conserving the happiness of the people, in securing love, in ensuring life, in bringing good and in warding away evil. When the mòthan is used as a love-philtre, the woman who gives it goes upon her left knee and plucks nine roots of the plant and knots them together, forming them into a cuach (ring). The woman places the ring in the mouth of the girl for whom it is made, in the name of the King of the sun and of the moon and of the stars, and in name of the Holy Three. When the girl meets her lover or a man whom she loves and whose love she desires to secure, she puts the ring in her mouth. And should the man kiss the girl when the mòthan is in her mouth [he] becomes henceforth her bondsman, bound to her everlastingly in cords infinitely finer than the gossamer net of the spider, and infinitely stronger than the adamant chain of the giant ... [It is] placed under parturient women to ensure delivery, and it is carried by wayfarers to safeguard them on their journeys. It is sewn by women in their bodice, and by men in their vest under the left arm ... To drink the milk of an animal that ate the mòthan ensures immunity from harm. If a man makes a miraculous escape it is said of him 'He drank the milk of the guileless cow that ate the mòthan.'

In contrast, the plant also has an entirely undeserved bad reputation. It is one of several plants called 'sheep-rot' in Scots, because they were thought to cause the disease in sheep. In reality, it was probably the waterlogged habitats in which the plant was found that caused the rot in sheep's hooves and feet. If you graze livestock in a bog, what can you expect?

Left: Butterwort by Linda Russell, watercolour (contemporary).

Below: Scottish dock by Fran Thomas, watercolour (contemporary).

56 Chapter 2: Wetlands

Docks *Rumex* species
Polygonaceae (dock family)

English:	Species names include Scottish dock (*R. aquaticus*), Curled dock (*R. crispus*) and Broad-leaved dock (*R. obtusifolius*)
Scots:	Docken, Bulmint, Bulwand, Cushycows, Rantytanty, Red sank, Smari dock
Gaelic:	Copag is the general name for docks, with qualifiers for each species: Copag albannach (*R. aquaticus*), Copag chamagach (*R. crispus*), Copag leathann (*R. obtusifolius*)
	In the case of English and Gaelic, many of the names are modern constructs for known species that would not traditionally have been distinguished in their uses.
Flowering:	Varies by species, but typically May to August
Habitat:	*Rumex aquaticus* (depicted here) is found in a very few boggy sites near Loch Lomond, *R. obtusifolius* (broad-leaved dock) is a widespread weedy plant in nutrient-rich grasslands and urban areas, and *R. crispus* (curled dock) is a classic coastal species.

These usually robust perennial herbs grow up to about 80 cm, with large, green leathery leaves of often tinged with red. The indistinct flowers are held on very dense racemes, often producing three-angled triangular fruits.

Medicinally, most species of dock were not distinguished from one another, so any would have been used in that most ubiquitous bit of hedge medicine – treating nettle stings. The roots were pounded up and used to treat bee stings and as a skin-tonic. The roots were also mixed with vinegar and lard to make an ointment for treating burns and scalds (similar recipes call for beeswax and butter). A tea or distillate made from dock was taken for scurvy.

Like other large-leaved, hairless plants, dock leaves were used to wrap butter.

Traditionally, the last sheaf of corn cut from the harvest was made into a corn dolly, dressed in dock and ragweed (*Senecio vulgaris*). This was then given to a rival farmer whose harvest was not yet in, thereby cursing them with the evil influence of the Cailleach, a hag who would sap the productivity from the land over the coming year.

In Shetland, where willow, hazel and other mainland basket-making materials were unavailable, dock stems were useful for providing rigidity. Fish baskets and traps, kishies (baskets for carrying for peat, manure and produce), brushes, mattresses, ropes, winnowing mats and grain tubs were all made from dock stems.

In dyeing, the roots were used as a mordant, and all parts of the plant helped to produce a range of olours. The leaves produce green-grey dye with alum and iron as mordant, and yellow-green with alum and cream of tartar. The roots give a salmon pink or greenish-brown dye with alum and iron and yellow-green with alum and cream of tartar. The dried fruits make a fawn dye with alum.

Willows *Salix* species
Salicaceae (the willow family)

Scots:	Sauch, Sauch buss, Sauch-willie, Sauch wand, Sauch tree, Scob, Widdie
Gaelic:	Seileach Many species have specific names in both Scots and Gaelic.
Flowering:	March to June, depending on the species
Habitat:	Most medium-sized and larger species grow in wet habitats. Most of the smaller shrubby species are restricted to upland mountain areas.

Willows are trees and shrubs of many sizes, ranging from 25 m for *Salix fragilis*, to just 10 cm for *Salix herbacea*. All are deciduous and have simple leaves with fine teeth on the margins. The flowers are catkins, which are present on both separate male and female plants. The fruit is a capsule producing tiny fluffy seeds.

Willow wood was used for making barrels and other purposes for which its fairly tough but pliant nature could be best put to use. The classic uses for willow are in basketwork and pain relief.

Scottish practices in basketry were much the same as elsewhere in Europe that used willow, with short-rotation coppice providing many whippy pliant stems for creels, screens and hurdles.

Although willow bark has not been used as an aspirin source for many years, we have an excellent account from the late 19th century held in the ethnobotanical archive at the Royal Botanic Gardens, Kew. This comes from a letter to a Kew employee from John Cowper, Secretary Board of Agriculture, dated 19 January 1909.

Many years ago, the chemical works at Beaverbank, Edinburgh, extracted bitters from the willow bark. After extracting they dried the bark and burned it. The smell of the burning bark was so nauseous that those living in the neighbourhood complained and the authorities put a stop to the burning of it. After that a market gardener named Black got it to put between the rows of strawberries instead of straw, but it did not suit the purpose and he gave it up. The next thing tried with it was the bunching together of market garden produce instead of straw ropes. It was found to answer the purpose so admirably that now all the market gardeners around Edinburgh use nothing else for bunching leeks, celery, rhubarb, broccoli, cauliflower, young turnips, young carrots, young beetroot etc. The chemical works at Beaverbank in order to get rid of it gave it away for years to all who chose to send for it. For many years that branch of the business has been stopped so that no bark could be got. The Royal Blind Asylum took up the trade to supply bark for the purpose already mentioned. I suppose they bring it from abroad. It comes into Leith and is sold at £3.15.0 per ton at ship side and is bought by market gardeners. The bark is bone dry and keeps useful for any length of time but it requires to be wet before using it. I can see we Scotch folk are more plodding, more persevering and far-searching than you gentlemen are across the border.

Other occasional uses for willow were in dyeing (for yellows and browns mostly) and charcoal production, but in both cases other trees were generally preferred.

Left: *Salix viminalis* from Sowerby's *English Botany*, Vol. 27 (1808).
Right: *Salix purpurea* from Sowerby's *English Botany*, Vol. 20 (1805).

Figwort *Scrophularia nodosa*
Scrophulariaceae (the figwort family)

English:	Rose-noble, Stinking Roger
Gaelic:	Lus-nan cnapan, Torranan (although see discussion on opposite page)
Flowering:	Mid to later summer (June to September)
Habitat:	Typically by streams or other bodies of water in lightly shaded woodlands

Below and right: Preserved specimen of figwort, RBGE Herbarium, http://data.rbge.org.uk/herb/E00655521.

This upright perennial herb has a cluster of tapering storage roots, and stems that grow to around 1.5 m tall. The simple, toothed leaves are lanceolate. The flowers are held in a tall spire-like raceme, and the individual flowers are short, purple-black bladders (often lighter inside), with red, green and yellow flushes.

Throughout Europe, figwort was used in the treatment of scrofula (a glandular form of tuberculosis), typically in the form of an infusion, and as an ointment for the treatment of piles. Lightfoot (1777) noted that it had fallen out of favour by his time in the later 1700s. Figwort appears in several spoken charms in Alexander Carmichael's *Carmina Gadelica*, although his identification of the plant is questionable. The description of the 'torranan' plant he identified as figwort, in one of his characteristically flowery tracts, sounds far more like a member of the carrot family:

> In the islands [figwort] was placed on the cow fetter, under the milk boyne, and over the barn door, to ensure milk in the cows. Having intoned the incantation of the torranan, the reciter said: 'The torranan is a blessed plant. It grows in sight of the sea. Its root is a cluster of four bulbs like the four teats of a cow. The stalk of the plant is as long as the arm, and the bloom is as large as the breast of a woman, and as pure white as the driven snow of the hill. It is full of the milk of grace and goodness and the gift of peace and power, and fills with the filling and ebbs with the ebbing tide.'

This description certainly does not fit with the spire-like raceme of purple-black flowers found on figwort; plants such as *Angelica sylvestris* (wild Angelica), *Ligusticum scoticum* (Scots lovage) or even the poisonous *Oenanthe crocata* or *O. lachenalii* (water-dropworts) fit the description and habitat better. It may be that one of his informants described the wrong plant, or sent the wrong species for identification to the Botanic Garden in Edinburgh. Either way, it is an intriguing botanical mystery.

Carmichael also explains that the plant held a more mystical reputation on the islands, but on the mainland the use was more prosaically medicinal: "the leaf is applied to cuts and bruises, and the tuber to sores and tumours".

Chapter 2: Wetlands **61**

62 Chapter 2: Wetlands

This group of several species of moss are compact (although relatively tall for mosses), are very bushy towards the tip, and have a whorled architecture through the whole plant. The leaves are tiny (1–4 mm), triangular, and densely arranged on the branchlets – one of many adaptations that allow them to absorb and retain water readily. Several species redden in autumn and winter, and many become a skeletal white when dry. As a moss it lacks flowers, but the sporangia are usually tipped with a globular black capsule.

Lightfoot (1777) recognised the importance of *Sphagnum* in forming peat, which was the staple fuel source across vast tracts of Scotland. The most famous uses for the whole plant exploit its absorbance: its specialised hyaline cells, closely packed leaves and compact habit make it an ideal absorber of pus, blood and other bodily fluids. It has been used as nappies, sanitary towels, wound dressings and liners for babies' cribs. When dry, it was thought to be mildly antiseptic and, indeed, it has been shown that many mosses, including *Sphagnum*, contain antimicrobial secondary metabolites and that the acid environment they create is tough for microbes to survive in.

The wad of *Sphagnum* packed against the chest of a corpse found at a Bronze-Age burial site at Ashgrove in Fife has been suggested to be a wound dressing. As with interpreting all archaeological finds, a degree of speculation is required, but certainly the practice has a lengthy recorded history. When cotton dressings were becoming scarce during the Napoleonic Wars, Highland soldiers in the British army instigated the use of *Sphagnum* dressings, and it was also used as the basis for wound dressings as recently as the World Wars. In fact, it is estimated that *Sphagnum* dressing production during World War I was around half a million kilograms per month (approximately 1 million dressings).

The bog mosses
Sphagnum species

Sphagnaceae (the bog-moss family)

Scots:	Flow moss, White peat
Gaelic:	Coinneach dearg, Fionnlach, Mointeach liath
Habitat:	These plants are responsible for peat formation, so they are the key group in peat-rich habitats. However, they will also grow readily anywhere there is enough regular or shallow standing water.

One of the major sources was the Highlands. During this period, collection was well organised. The moss was collected manually, with any convenient implement, such as rakes or garden forks, being used to help. To expel at least some of the water the plants contained, the *Sphagnum* was then squeezed by hand or put into sacks then jumped on. Other plant matter was then removed from the moss before it was carried by cart to the nearest railway for shipment to processing plants. Here, it was sterilised, packed into cakes and then transported for further processing into dressings. Maud Grieve (1931) gives a good account of the process.

Sphagnum is said to be more absorbent, and absorb liquids more uniformly throughout its surface, than cotton dressings. It was also reported as having the added benefit of being less irritating to open wounds than more typical cotton dressings. Despite this, the use of *Sphagnum* dressings died out after the end of the wars, when cotton was no longer at a premium.

Left: *Sphagnum* bog moss by Fran Thomas, watercolour (contemporary).

Shetland monkeyflower: a new find

This local variant of monkeyflower (*Mimulus guttatus*) is a fascinating example of plant diversity and the impact of humanity. Native to North America, its ancestors were introduced to the UK as ornamentals several centuries ago, and escaped cultivation to establish populations in the wild. This variant was found on Shetland, near Quarff, south of Lerwick, by a research team from the University of Stirling. Its genetics were studied by Violeta Simon-Porcar, and showed an interesting phenomenon.

The monkeyflower group is prone to hybridising (crossbreeding between species) and multiplying the number of genes they have (a phenomenon called polyploidy). In this case, it was the latter. The resulting increase in chromosome numbers caused the plants to be larger, with larger leaves, thicker stems and bigger flowers. RBGE is studying similar phenomena in Scotland and worldwide to help understand the processes that lead to new biodiversity.

Left: Shetland monkeyflower by Janet Watson, watercolour (contemporary).

Minor species of wetlands and waterways

Wetlands harbour many specialised plants – fully aquatic species can be adapted to still or fast-flowing water, whereas a series of others prefer differing degrees of wet. Perhaps because of their distinctive habitat and often striking adaptations to the environment, wetland plants are generally easy to identify. This, in turn, may make these plants more commonly used, because it is always better to use a plant you can identify easily rather than confuse it with something more dangerous. There are some exceptions, however.

Caltha palustris is a striking, sturdy member of the buttercup family with large, dish-shaped flowers. Although it is most commonly known as marsh marigold, it is one of the plants in Scotland with the highest number of common names (more than 25 different local names and variants are known for this one species alone). Oddly, however, it was little used; the more common buttercups from drier ground seem to have been preferred for use in traditional medicine.

Another waterside speciality is water avens (*Geum rivale*). This is one of two native avens. The other is *Geum urbanum* (wood avens), which is a common plant in town parks and gardens, hence the taxonomic name. The two species hybridise to produce many intermediate forms. Much like the marsh marigold, water avens is mostly used as an ornamental plant, but the root of wood avens has a warm spicy smell, and was occasionally used as a flavouring in beers and liqueurs. For this reason, it is sometimes known as clove-root.

Above: Marsh marigold by Sandra Russell, watercolour (contemporary).

Butterbur (*Petasites hybridus*) is one of several species in the daisy family that flower before the leaves come up, giving them the name 'son-before-father'. The large leaves look very like rhubarb, and can be found lining the edges of native streams from late April. As the name suggests, the large, soft leaves were used to wrap butter. It was also one of the key ingredients in a traditional love charm.

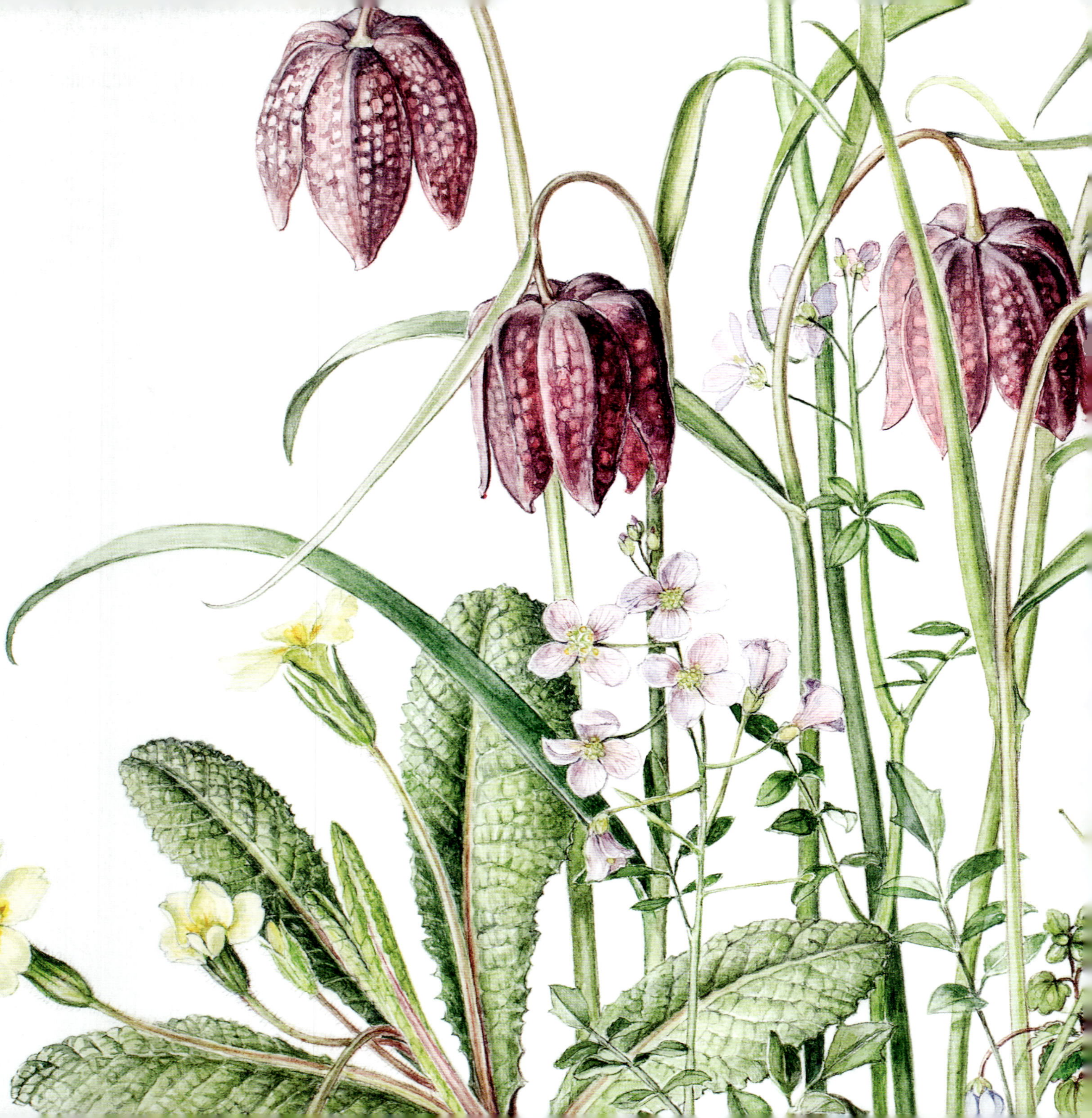

Chapter 3
Grasslands

Scotland's grassland communities are far more varied than they might appear at first sight. As with any other habitat, the soil, water availability, nutrients and sunlight all define the interplay of species and the niches available to the plants, animals and other organisms that survive there. From coastal lowlands, through urban areas, farmland field margins, and open areas by woods, heaths and mountainsides, all have their own variations on the grassland theme. Of course, grasslands are dominated by grass species, but they also harbour a wide variety of other herbaceous plants with more conspicuous flowers, as well as a range of more shrubby species. This rich tapestry of habitat types and species diversity gave humans a versatile resource beyond grazing land. Grasslands were just as important as coastal areas in the domestic economy, medicine chest and larder of the past.

Paradoxically, nutrient-rich 'improved' grasslands are often the least biodiverse. In places where we have added fertilisers, somewhat thuggish and competitive plants thrive, edging out rarer and less robust species. In upland areas, acid grasslands blend into heath and woodlands, sharing many species with them but also harbouring their own particular species. The machair habitats of the Atlantic coast and islands are particular gems and a real Scottish speciality, home to some beautiful orchids and a wide variety of other herbaceous delights, including an array of versatile dye plants.

Left: The non-native, but commonly grown, snakeshead fritillary (*Fritillaria meleagris*) is often naturalised in urban grasslands and gardens. It was an iconic subject of Charles Rennie-Macintosh. This watercolour by Janet Dyer shows a composite study of snakeshead fritillary growing among cuckooflower (of wet meadows), primrose (of woodland clearings) and speedwell (from urban grasslands).

Above and right: Preserved specimen of Yarrow, RBGE Herbarium, http://data.rbge.org.uk/herb/E00903214.

Yarrow *Achillea millefolium*

Compositae, Asteraceae (daisy family)

English:	Hundred-leaved grass, Thousand-leaved clover
Scots:	Doggie's brose, Meal and folie, Melancholy, Moleery tea, Stanch-girse, Yarra
Gaelic:	Earr thalmainn
Flowering:	July to September
Habitat:	Grasslands and lawns; copes very well with close grazing and being walked over.

This perennial herb with distinctive feathery leaves grows up to 50 cm high, but usually to 20–30 cm and often much less when grazed or mowed. The flowers are a dense cluster of tiny silvery-white and pink flowers in a daisy-like head called a capitulum, giving the impression of a larger flower (to about 1 cm across). These capitula are then grouped into flat heads that act as landing pads for pollinators.

Yarrow was traditionally used as a wound-healer. Lightfoot discusses this in *Flora Scotica* (1777):

> The plant has an astringent quality, and is reckon'd good to stop all kinds of hoemorrhagies, and to heal wounds, but is out of use in the present practice.

The Highlanders still continue to make an ointment of it to heal and dry up wounds.

Lightfoot also explains that a leaf stuck up the nostril was used to *cause* nosebleeds, thereby helping to alleviate headaches, but writing only a few years earlier, James Robertson (1767–1771) explains how a decoction warmed and applied with a feather to the nose was used to *stop* nosebleeds. Writing in 1931, Maud Grieve explains that an ointment made in the Highlands from yarrow was used for piles, and to treat scab in sheep.

As with many plants, it yields a pale-yellow dye with alum and cream of tartar as mordant.

Mary Beith's *Healing Threads* (2004) explains that yarrow was used in Gaelic-speaking areas to help divine the success of a relationship. Cut before sunrise and placed under a pillow, it would grant girls the ability to dream of their sweethearts. If the man was facing them in the dream, the pair would marry, but if he had his back turned, they would not.

Alexander Carmichael (1900) describes a number of charms and poems relating to yarrow. The following was to be said while picking the plant:

> I will pluck the yarrow fair, That more benign will be my face, That more warm shall be my lips, That more chaste shall be my speech, Be my speech the beams of the sun, Be my lips the sap of the strawberry. May I be an isle in the sea, May I be a hill on the shore, May I be a star in the waning of the moon, May I be a staff to the weak, Wound can I every man, Wound can no man me.

Perhaps the romance of such writing is a little deflated by the knowledge that the plant was also used to make pile ointment!

Chapter 3: Grasslands 69

Broom *Cytisus scoparius*
Leguminosae, Fabaceae (pea family)

Scots:	Breem, Brome, Broume, Brume
Gaelic:	Bealaidh
Flowering:	June and July
Habitat:	Scrubland and forest margins. Broom tolerates a wide range of disturbed conditions, and often prefers drier soils.

Broom is one of our largest pea family members. It is a golden-flowered shrub distinguished from gorse by its tiny trefoil leaves and lack of either spines or a coconut scent. The stems are flexible, upright and dark green, and do most of the photosynthesis. As a member of the Genisteae group of legumes, it contains a plethora of interesting compounds.

The pea flowers are around 2 cm long, and bright golden yellow. The fruits, at 4–6 cm long, are dark brown pea pods that on warm days split and fire the seeds across several metres with an audible pop.

Being tough and fibrous, the branches of broom were, of course, used for making the switch of brooms and other brushes. These properties also made them a durable thatch, although finding enough for a whole roof could be a challenge.

Broom is a versatile dye plant; the flowers give shades of yellow from pale cream to gold, depending on the mordant. The young branches produce a green colour.

Despite the plant's toxicity, the tops of broom were used for flavouring a barley-based beer. Lightfoot (1777) explains they were used in Breadalbane and Ross-shire to produce a fermented drink in its own right, and this broom wine is said to have a hint of almond in the flavouring. Dr George Johnston (1853) states that one of these brews "exerts an intoxicating influence on man", and quotes the Scottish poet Allan Ramsay's address to a landlady notorious for her powerful brew:

> Some said it was the pith o' broom,
> That she stow'd in her masking-loom,
> Which in our heads rais'd sic a soom;
> Or some wild seed,
> Which aft the chaping stoup did toom,
> But fill'd our head.

The young petals are thought to be relatively inert so are often nibbled by impromptu foragers, and were once used to produce a tea. The seeds were also removed from the pods and roasted to make a good substitute for coffee, although this was more typically practised in continental Europe.

The plant also had medicinal uses. In Fife, the miners used broom tops and nettles infused in water as a treatment for dropsy. Mary Beith (2004) explains that the flowers or seeds were used to induce vomiting and that nosebleeds were treated by tying a bunch of the plant around the patient's neck, although it is difficult to imagine how this was supposed to have worked.

For some reason, records for broom often mention where in Scotland a particular use occurred. It is relatively uncommon to have such specific detail.

Below: Field illustration of broom by Jane Wisely, watercolour (1938).

Chapter 3: Grasslands 71

72 Chapter 3: Grasslands

Lady's bedstraw *Galium verum*
Rubiaceae (coffee family)

English:	Lady's bed
Scots:	Keeslip, Ruin
Gaelic:	Lus an leasaich
Flowering:	June to August
Habitat:	Open grassland. It does best in well-drained, sunny spots.

This dark-green perennial herb grows to 50 cm tall. The slender, soft, needle-like leaves grow in whorls of eight to twelve up the stem, which is topped by a panicle of small flowers with four yellow petals.

Galium verum is a useful plant but was once collected in such quantities as to threaten it and its habitats. The tops were used to make the straw that gives the plant its common English name and one of its Gaelic names, *Leabadh ban-sith*, which can be read as the beautifully evocative 'banshee's bed'.

The roots produce a good red dye mentioned by many writers, but the particularly observant James Robertson, writing in the 1760s explains this as follows:

> From a decoction of the roots of Galium verum/Yellow Ladies Bed straw, a carmine dye is prepared, equal to that produced from the roots of Madder. As the Galium grows however amongst the sand, the people are not allowed to dig it.

Although it gives good colour, the roots are tiny and many are needed, which necessitates a lot of work for their collection. This explains the significant ecological damage caused by pulling them from the sandy soils. As the soils destabilised and were blown or washed away, the buffer zones provided by the machair grasslands were lost. This was a particular problem on these sandy soils of the west coast and islands. Collection had been banned on and off since the late 16th century. The aerial parts yield a yellow dye.

On Arran and some of the Western Isles, *Galium verum* was used as a decoction (often with nettles added) to curdle milk, although which part is not specified.

Left: Field illustration of lady's bedstraw by Jane Wisely, watercolour (1968). The leaves have been stylised to look more feathery than in real life.

Bitter vetch *Lathyrus linifolius*
Leguminosae/Fabaceae (pea family)

English:	Bitter vetch, Heath pea, Tuberous pea, Tuberous vetch
Scots:	Caperoiles, Carmele, Heath pease, Heather-pease, Heath-vetch, Knaphard, Knapperts, Napple, Wood-pease
Gaelic:	Cairt leamhna
Flowering:	May to August
Habitat:	Frequent, in acid grassland in uplands, and often in heath

Below: Bitter vetch by Sarah Roberts, watercolour (contemporary).

This perennial herb has running rhizomes with tubers at some nodes. Its aerial stems grow to 30 cm. The leaves have a small point at the tip rather than the tendril typical for many of its relatives, with between two and four pairs of leaflets. There are around five flowers on each flower head. These form short, few-seeded pods, each 1.5 cm long and pea-like, with purply-pink petals.

Bitter vetch gives us a fascinating example of traditional plant use in Scotland, around which a whole minor mythology has been built. Many tuberous plants have been nibbled, chewed or brewed throughout Europe, but this species is often seen as characteristically Scottish in its use. One of our earliest reliable records of bitter vetch comes from Martin Martin (1703), when he collected it on his travels throughout Scotland on behalf of Robert Sibbald.

Carmel, alias knaphard, by Mr. James Sutherland called argatilis sylvaticus. It has a blue flower in July. The plant itself is not used, but the root is eaten to expel wind and they say it prevents drunkenness by frequent chewing of it; and being so used, gives a good relish to all liquors, milk only excepted. It is aromatic and the natives prefer it to spice for brewing aquavitae. The root will keep for many years; some say that it is cordial, and allays hunger.

Intrigued by the plant, Sibbald devotes a special section in his *Provision for the Poor in Time of Dearth and Scarcity* to it, and it is illustrated in his natural history of Scotland, *Scotia Illustrata sive Prodromus Historia Naturalis*.

Several writers in the mid- and late-1700s followed, explaining the various uses of bitter vetch as an appetite suppressant, seeming hangover cure and flavouring for alcohol. Thomas Pennant (1774) explains that infused in whisky, it made an "agreeable beverage and, like the Nepenthe of the Greeks, exhilarates the mind". Lightfoot, another non-Gaelic-speaker, wrote about it three years later:

> The Highlanders have great esteem for the tubercles of the roots of this plant; they dry and chew them in general to give better relish to their liquor; they also affirm them to be good against most disorders of the thorax, and that by the use of them they are able to repel hunger and thirst for a long time. In Breadalbane and Rosshire they sometimes bruise and steep them in water and make an agreeable fermented liquor with them. They have a sweet taste, something like the root of liquorice, and when boiled, we are told, are well flavoured and nutrative, and in times of scarcity have served as a substitute for bread.

James Robertson reported that "it is said to be aromatic and is eaten before drinking strong liquors to prevent intoxication", and on Skye it is eaten when people feel thirsty or faint.

The descriptions of chewing the plant to counter fatigue are reminiscent of betel in south Asia, khat in the Arabian peninsula and the Horn of Africa, and coca in Andean South America.

Below: Bitter vetch from *Sibbald's Prodromus* (1684).

Bistorts *Persicaria* species
Polygonaceae (dock family)

English:	Snakeroot, Snakeweed, Spotted arsmart (*P. maculosa*), Hot arsmart, Spotted knotweed, Water-pepper (*P. hydropiper*)
Scots:	Yallowin' girse (bistorts generally); Pencuir kale (*P. bistorta*); Deid arsmart, Flooreing soorick, Spotted arsmart, Useless (*P. maculosa*)
Gaelic:	Lusan-glùineach (*P. amphibia*), Lus chrann ceusaidh (*P. maculosa*), Glùineach theth (*P. hydropiper*)
Habitats:	Various grasslands, although mostly wet meadows. Completely aquatic in the case of *P. amphibia*.

The leaves are long-ovate or lanceolate, sometimes with dark markings on them. The base of the leaf has a papery sheath around the stem; this is called an ochrea. The stems are sometimes red or pink, and the flowers are small, borne on upright spikes, and various shades of maroon, through pink to white. The fruits are dry.

The bistorts are a group of related species mostly in the genus *Persicaria*. *Persicaria bistorta* was boiled and eaten as greens, and James Robertson (1760s) mentions that the poor of Deeside, where small bistort (*P. vivipara*) was abundant, dried the roots and seeds and ground them into a meal, which was considered "wholesome and nourishing". Lightfoot (1777) explains that the root of bistort had an acid taste and was used as a powerful astringent, while Mary Beith (2004) mentions its alleged efficacy in treating urinary complaints. Lightfoot also mentions this use for water pepper, *P. hydropiper* (as a diuretic in particular), although he explains it was little used even in his day.

Members of the group were widely and generally used as a source of yellow dyes, and for this reason in Shetland were known as yallowin' girse. Gaelic names for redshank (*P. maculosa*) include Lus chrann ceusaidh, an allusion to the belief it grew from the blood of Christ on the crucifix, or from the cross itself. Another name, am boinne fola (blood-spot), relates to the same.

One of Robertson's accounts of small bistort is a fascinating insight into a recurring theme in the works of many 17th and 18th century writers, showing the dichotomy between 'educated' urban lowlanders and superstitious Highlanders:

> The inhabitants of this tract [near Invereichy], like several of their ignorant Brethern in the highlands, believe that some people tho' at a distance have the power of affecting milk in such a manner that it will yield neither butter nor cheese. Others are believed to possess the power of undoing the spell, and they effect their purpose by putting a few leaves of the small Bistort, Milkwort and yellow Devil's bit into a rag which is tied and immersed into the bewitched milk.

Right: A field illustration of a bistort by Jane Wisely, watercolour (1940), showing the distinctive papery ochrea at the nodes.

Chapter 3: Grasslands 77

78 Chapter 3: Grasslands

Silverweed *Potentilla anserina*
Rosaceae (rose family)

English:	Dog Tansy, Fair days, Goose grass, Mascorn, Moor grass, Moss crop, Moss grass, Swine beads, Swine's grass
Scots:	Fair girse, Swine's murricks (Pig's roots), Wild skirret
Gaelic:	Brisgean
Flowering:	June to August
Habitat:	Low, usually well-drained grassland on sandy soils. The plant is very tolerant of trampling, so is commonly found at the edge of paths, although the really large specimens are in ungrazed corners of machair or particularly rich soils.

This perennial herb grows to 20 cm, but with a spread of bright red runners creeping up to 50 cm. The feathery compound leaves are very distinctive with dense silvery hairs – giving the plant its main common English name of silverweed. The flowers are solitary and around 2 cm across; typical for most cinquefoils, they have five petals, which are bright yellow in this species.

Silverweed is *the* classic famine foods in Scotland. Lightfoot (1777) explains that the roots taste like parsnips and were typically boiled or roasted. Roasting was generally better for retaining the flavour. Mary Beith (2004) suggests this was one of the main starchy root crops prior to the introduction of the potato, and it is possible that active cultivation was happening both before and after the potato was introduced to Europe. Lightfoot certainly suggests they were a true staple in Coll and Tiree. There have been occasional searches in more recent times for fabled spots where plants with monster tubers might have grown, but for conservation reasons these are not to be encouraged, and besides have not met with much success. There is also some speculation (although it is really only speculation) that conscious selection of plants with larger tubers may have been happening; essentially, people may have been developing silverweed as a crop. Much of this speculation can be traced to Carmichael's account in *Carmina Gadelica*:

> Certain places are remembered for the cultivation of silverweed. One of these was Lag nam Tanchasg in Paible, North Uist, where a man could sustain himself on a square of ground his own length. In dividing common ground, the people lotted their land for brisgein much as they lotted their land for fishing banks at sea and their fish on shore. The poorer people exchanged brisgein with the richer for corn and meal, quantity for quantity and quality for quality. The brisgein was sometimes boiled in pots, sometimes roasted on stoves and sometimes dried and ground into meal for bread and porridge. It was considered palatable and nutritious.

Silverweed has a few other uses: with alum as a mordant it can be used to produce a pale-yellow dye, and when infused in buttermilk was used to make the skin paler by removing tanning and freckles.

Left: Field illustration of silverweed by Jane Wisely, watercolour (1940).

Tormentil *Potentilla erecta*
Rosaceae (rose family)

English:	Blood-root, Flesh and blood, Shepherd's knot
Scots:	Aert-bark, Eart-bart, Ewe-daisy
Gaelic:	Cairt-láir
Flowering:	June to September
Habitat:	Acid grassland blending into heather

This perennial herb grows to 30 cm tall. It is often many-stemmed and forms a tangled mass or runs through heather and grass. The 'leaves' appear palmate with five 'fingers', each with teeth on the tips. Each 'leaf' is actually a three-part leaf and two stipules at the base. The flowers are about 1 cm across, and usually have four yellow petals.

The rhizome can grow to the size of new potato, has dark-brown skin, and was the most useful part of this versatile plant. The red-tinged dense flesh is rich in tannin. Because tormentil was so plentiful, it was a major source of tannin in areas without ready access to tree bark or imported tannin. The practice of obtaining tannin from tormentil was widespread throughout Scotland, and particularly common in the Western and Northern Isles. Lightfoot and Robertson, writing in the 1770s, both mention the process. The tubers were boiled in water, then the leather, rope or fishing nets being tanned were immersed in the liquid when it was cool. Collection was onerous: although the tuber is easily identified, one record explains that a man would work almost a whole day to dig up enough plants to prepare a single decoction. Robertson explains that the inhabitants of Skye preferred tubers from rough, remote ground, because it was thought to be three times as effective as plants growing on cultivated or pasture land. This is an intriguing observation that may reflect a difference in the biochemistry of the plants in different growing conditions.

During Lightfoot's time, plants were overharvested in Coll and Tiree, leading to a local ban on collecting. Although this near-industrial-scale use was the main one, tormentil was a versatile plant and the tubers were also a source of a red dye.

Medicinally, a decoction of the tuber in milk was used to treat stomach complaints, with specific mentions of dysentery and diarrhoea, as well as killing off gastro-intestinal parasites such as worms. These were evidently widespread and long-lasting practices from the borders to Shetland and the Western Isles, with records throughout the 17th to the 19th centuries. An alcoholic distillate was used to strengthen the teeth and gums of scurvy patients, and until recent times it was chewed to heal sore lips. The tannin presumably gives any medicinal preparations an astringent taste, so it would have been used whenever astringents were called for. Passing mentions suggest it was also used as a medicine for cattle, certainly in the later 1800s, but without much detail on what was being treated or how.

Right: Field illustration of tormentil by Jane Wisely, watercolour (1936).

Chapter 3: Grasslands 81

Sorrel and sheep's sorrell
Rumex acetosa / R. acetosella

Polygonaceae

Many of the common names apply to both of these species, and sometimes to other members of the Polygonaceae (*Rumex*, *Polygonum* and *Bistorta*)

English:	Red shank, Sour dock
Scots:	Lamb soorocks/sourocks, Rabbit's sugar, Ranti-tantie, Reid shank, Soorocks (and many variant spellings), Soordock, Sookie soorocks, Soor leek
Gaelic:	Samh (*R. acetosa*), Sealbhag nan Caorach (*R. acetosella*)
Flowering:	June to August
Habitat:	Grasslands: from rich agricultural soils, and well-drained dry habitats to wetter meadows. The smaller *R. acetosella* is common on well-drained, rocky ground.

Sorrel (*Rumex acetosa*) is a perennial herb that is hairless and has glossy, slightly fleshy leaves. It grows to a maximum of 80 cm, but usually shorter than this. The leaves are shaped like an arrow head (sagittate), with a blade about 10–15 cm long and backward-pointing tines. On the much smaller sheep's sorrel (*R. acetosella*), the tines point out towards the side (hastate) and the plant is often-red-tinged throughout. The flowers on both species are borne on an upright raceme, and do not have obvious petals (they are present, but modified into little wings). The flowers and subsequent triangular fruits are green or various shades of red.

Sorrel has long been cultivated, semi-cultivated and gathered from the wild as a pot herb and salad vegetable. It generally grows very well, so was a common medicinal plant in monasteries and later physic gardens. The crossover between food and medicine is nicely illustrated by the following from Martin Martin, writing at the end of the 17th century:

> The men [of St Kilda] are stronger than the inhabitants of the opposite Western Isles; they feed much on fowl, especially the solan geese [gannet, *Sula bassanica*], puffin [*Fratercula arctica*] and fulmar [*Fulmarus glacialis*], eating no salt with them. This is believed to be the cause of a leprosy that has broken out among them of late. One of them that was become corpulent, and had his throat almost shut up, being advised by me to take salt with his meat, to exercise himself more in the fields than he had done of late, to forbear eating the fat of the fowl, and the fat pudding called giben, and to eat sorrel, was very much concerned because all this was very disagreeable; and my advising him to eat sorrel was perfectly a surprise to him; but when I bid him consider how the fat fulmar eat this plant he was at last disposed to take my advice; and by this means alone in few days after, his voice was much clearer, his appetite recovered, and he was in a fair way of recovery.

The distinctive acid freshness made it a popular flavouring – a use which continues today. Although it was used to treat scurvy in Lightfoot's time, the main source of the flavour is oxalic acid, rather than ascorbic acid (vitamin C). Like all plants, it does contain vitamin C, however, so it may have been effective. Much modern literature suggests that too much oxalate can cause kidney damage so it should only be eaten in moderation. Cooking with mature leaves helps to minimise the oxalate concentration, so researchers Tuazon-Nertea and Savage (2013) recommend its use in pesto and soup.

The whole plant gives a yellow dye and can be used as an acid mordant for other dyestuffs. The roots were traditionally used for a red dye.

Left: Sorrel from Sowerby's *English Botany*, Vol. 2 (1793).

84 Chapter 3: Grasslands

Gorse *Ulex europaeus*
Leguminosae/Fabaceae (pea family)

English:	Whins, Whin-cow
Scots:	Carlin-spurs, Carling spurs, Fun, Fun bus, Furze, Whun, Whun bush
Gaelic:	Conasg
Flowering:	The main flowering is in late spring to late summer, but gorse can flower all year round, hence the classic phrase 'When gorse is in flower, then kissing is in season'.
Habitat:	Hedgerows and woodland margins, especially by paths and roadsides

An unmistakable dark green, spiny shrub that grows to 2.5 m high. The leaves are reduced to spines, with side shoots acting as spines too. The plant is a member of the legume family, with nitrogen-fixing nodules on the roots; these seem to give it an advantage in nitrogen-poor soils. It prefers drier sites, and slopes, and can be very invasive even in its native range, and is a major problem in Australia, New Zealand and parts of the Americas. The flowers are vanilla- and coconut-scented pea flowers with golden-yellow petals. The flowers have a seemingly ingenious mechanism to burst a cloud of pollen over the first bee that visits. The fruits are a very short, fuzzy legume (pea pod), and the seeds are ejected forcefully from the dry pods for several metres when they burst, with an audible pop, on a hot day.

Left: Gorse by Sharon Tingey, watercolour (contemporary).

There are two other species of gorse in the UK, but they are rare or absent in Scotland.

The flowers were used as an edible garnish in salads and were sometimes used among other ingredients in distillates; for example, they are used in gins even today. The main use of gorse was as animal fodder. Plants would be harvested and dried, then ground up on a whinstone (a millstone specifically used for the purpose). The rock type known as whinstone takes its name from these millstones.

The young shoots produce a green dye and the bark and woodier twigs a brown dye. The flowers can give a wide range of colours, almost up to the same vivid gold as the flowers themselves. Paler yellows are achieved with alum and cream of tartar, gold if chrome is the mordant, and green with iron.

In Fife, gorse was said to be unlucky, particularly when presented as a gift to another person.

Above: Scots primrose by Kathy Pickles, watercolour (contemporary). The tiny *Primula scotica* is found only in Scotland. It grows in grassland at the tops of cliffs in Caithness and Orkney. It suffered terribly from over-collection in the past.

Minor species of grasslands

A wide range of species were used from grasslands, and the 17th and 18th century botanists have luckily recorded many of these uses. John Lightfoot (1777) and James Robertson (1768) explain how the inhabitants of Skye made ropes and nets from purple moor grass. Robertson describes how they were pulled up by the roots, beaten and dried and Lightfoot says, of two grassland ferns, adder's tongue (*Ophioglossum vulgatum*) and moonwort (*Botrychium lunaria*), that, "The common people sometimes make an ointment of the fresh leaves and use it as a vulnerary to green wounds" (i.e. fresh wounds).

This preparation was generally known as 'adder's spear ointment', and was probably widely used throughout Europe. The superficial similarity between adder's tongue and the head of a snake meant it had been used as a treatment for snake bites – an example of the Doctrine of Signatures. By Lightfoot's time, in the Enlightenment, adder's tongue was probably falling out of favour, although it still appears in *Pharmacopeia Edimburgensis*, one of the standard reference works on medicines of the 18th century. The plant is rare now, with modern fertiliser regimens and overgrazing probable contributing factors.

Other grassland plants, such as burnet saxifrage (*Pimpinella saxifraga*) and spignel (*Meum athamanticum*), may also have become rarer since the 18th century. Lightfoot and Robertson both mention the use of these two species as a carminative, summarised by Robertson thus:

> Self-heal (*Prunella vulgaris*) remains common, and was a versatile vulnerary – bruised and applied to fresh wounds, or mixed with goldenrod (*Solidago virgaurea*) and butter as an ointment. Infused, it was drunk in broth in cases of internal bleeding and injected (or used as an enema) to treat dysentery.

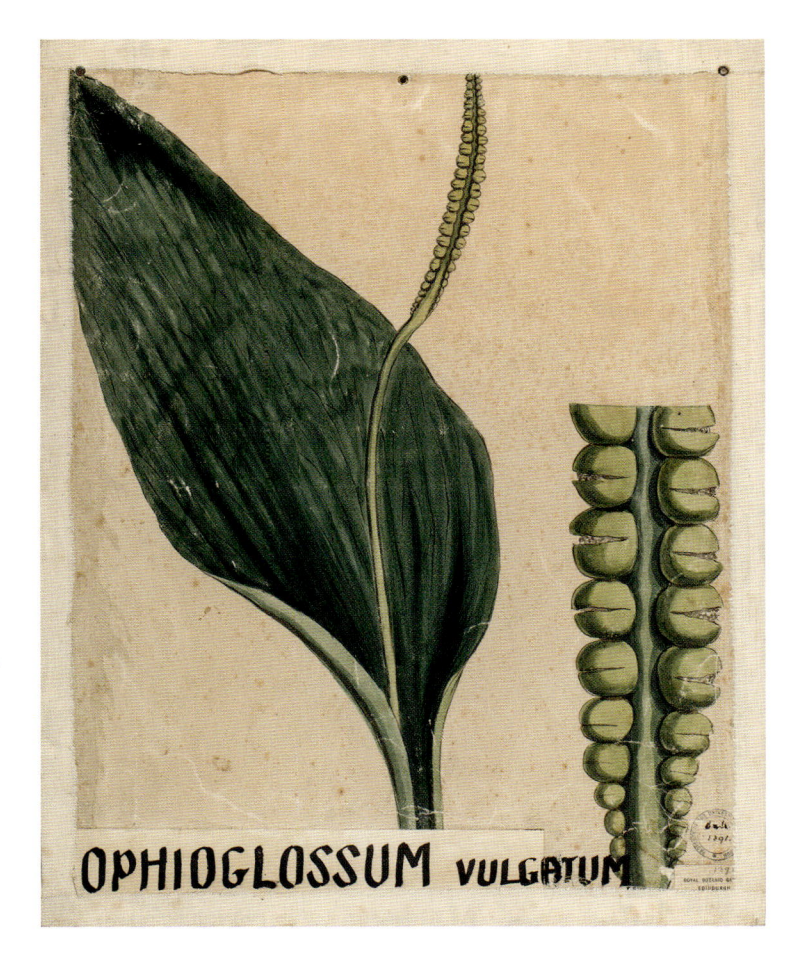

Above: Adder's tongue, John Hutton Balfour teaching diagram collection (19th century).

Below: Solomon's seal, Victoria Braithwaite, watercolour (contemporary). This species (*Polygonatum multiflorum*) is a rare plant from the south west of Scotland and debatably native.

Chapter 4
Woodlands

Woodlands and hedgerows harbour a huge diversity of life and, by many measures, include the most biodiverse habitats in Scotland. Through land clearance and harvesting for timber, charcoal, tannins and other non-timber forest products, they have been reduced to a fragment of their once extensive coverage, which may have extended to 80 per cent of Scotland around 6,000 years ago. Upland management for game undoubtedly has had a strong impact on regeneration, particularly for Scots Pine woodland, but the west coast Atlantic oakwoods are generally faring better. Intriguingly, research has shown that even these ancient-seeming primal woodlands have a long history of use by humans that can still be seen in the patterns of diversity of lichens and mosses.

In the central belt of Scotland, particularly near the large conurbations, the non-native sycamore (*Acer pseudoplatanus*) and beech (*Fagus sylvatica*) have both integrated with native woodlands to such a degree that they have been adopted by many non-native plants and animals as partners, hosts and food. This is fairly typical; many garden, parkland and estate escapees have become a natural part of the landscape in woodlands. Many of these were brought in as ornamentals or in the early days of forestry research, but a group called the archaeophytes have been here longer. A great example of this is sweet cicely (*Myrrhis odorata*), which was brought in from northern Europe, possibly through monastic gardens in the Middle Ages, although it is difficult to say for sure. This lovely woodland herb is quite distinctive, with softly hairy, finely dissected fern-like leaves. At the base of many leaflets is a patch of silvery white – an essential feature to help in its identification. The whole plant smells sweetly of aniseed when crushed, and it was used as a pot herb, salad and strewing herb.

Left: Lesser celandine, Janet Watson, watercolour (contemporary).

Woodland herbs

Woodland and hedgerows harbour many species of herbaceous plants; the varying soil conditions, paired with light levels that change through the seasons, in clearings and at the fringes of the wood, make for a heterogeneous habitat. This allows a wide diversity of species to thrive. Some of these have adapted to appear early in the year before the canopy of leaves shades out the sun. These give us beautiful displays of spring flowers. Others are specialised for shade tolerance and can be found lurking in some surprisingly dim corners. Many of these plants were used in similar ways across Europe, as pot herbs or as cures (often of dubious efficacy) under the widespread traditional medicine systems.

Arum maculatum (lords and ladies), although a distinctive species, with several names in Scots and Gaelic, was little used. The tubers were apparently used through northern Europe as an a emergency famine food, perhaps dried, roasted and milled to a flour, but the whole plant contains irritant calcium oxalate crystals. These make it dangerously unpleasant to eat, and many modern efforts to recreate the famine food arum flour have failed – so do not try!

Another toxic but distinctive little woodland plant is *Ficaria verna* (lesser celandine). Like many of its related buttercups, the sap can be an irritant, so was used throughout later medieval Europe to raise blisters to help 'balance the humours'. The sap was also used more prosaically to combat warts, a tradition that continued until the early 1900s on Colonsay at least. Lesser celandine also had a role in the Doctrine of Signatures, the widespread medicine system by which cures were informed by the appearance of the medicine. The storage roots of this cheery early spring flower are hideously suggestive of painful haemorrhoids, hence the name pilewort and its use in combatting the condition.

Honeysuckle (*Lonicera periclymenum*) may have been the 'woodbine' that magical protective hoops were made from, although ivy is a possibility too. Although the berries are inedible at best, and usually considered poisonous, they have been used to make a type of wine. Medicinally, they were used to treat asthma and bronchitis.

The use of toxic woodland plants seems almost to be a theme. Martin Martin (1703) refers to a plant called mercury, which may be *Mercurialis perennis* (dog's mercury), a particularly poisonous plant in the spurge family: "They make blisters of the plant mercury, and some of the vulgar use it as a purge, for which it serves both ways." In this case, he is referring to raising blisters, which were then lanced to allow the "ill humour to flow out". Foxglove is also known as wild mercury and Scotch mercury, so it may be that he was referring to this or another plant – a perpetual puzzle for the modern botanist.

After dysentery, or other diarrhoetic conditions, the astringent root of *Geum urbanum* (wood avens, or herb bennet) was infused in wine, whole or in powder form, and drunk. In several European countries the name is a variant of clove root, because the roots were used as a clove-like flavouring.

Herb robert (*Geranium robertianum*) was a versatile little plant. Mary Beith (2004) mentions its use in treatments for cancers, and the Gaelic plant-name lexicographer Cameron (writing at the end of the 19th century) explains that this species and bloody cranesbill (*G. sanguineum*) "have been and are held in great repute by the Highlanders, on account of their astringent and vulnerary properties." It is a smelly plant and somewhat reminiscent of burnt rubber when crushed, a property Lightfoot says was used to "drive away bugs".

The pleasantly scented *Galium odoratum* (sweet woodruff) was widely used in bedstraws, or as a strewing herb, but was also used to treat consumptions and other chest complaints. As Mary Beith 2004) explains, the Gaelic name lus na caithimh means wasting wort (reflecting its use to treat wasting diseases).

Primroses (*Primula vulgaris*) are mentioned in one of Alexander Carmichael's collected poems (1900) as a special spring treat for children to eat. The plants were used to treat minor cuts and grazes, burns, boils and abscesses – usually tied in place under a bandage.

Martin Martin recommends violets (*Viola* species) boiled in whey as a cooling and refreshing drink for feverish patients.

Among the classic pot herbs is *Alliaria petiolata* (garlic mustard) with its subtle garlic flavour. This member of the cabbage family remains popular among foragers today, and it commonly produces early and late flushes of leaves, resulting in a long cropping season. Robert Sibbald names it as one of more than 50 edible plants in his *Provision for the Poor in Time of Dearth and Scarcity*. This account also includes many other woodland species, such as *Conopodium majus* (pignut), the dense, nutty flesh of whose roughly spherical tubers has long been popular as a wild food, but care has to be taken not to mistake it for the superficially similar tubers of poisonous plants such as *Arum maculatum*.

Above and right:
Ramsons by Lyn Campbell, watercolour (contemporary).

Ramsons *Allium ursinum*
Amaryllidaceae (the snowdrop and garlic family)

English:	Wild leek
Scots:	Ramps
Gaelic:	Creamh
Flowering:	Late April to May
Habitat:	Grows in large drifts in woodlands, flowering as the canopy closes over; little is left by mid-summer.

This perennial bulbous herb has flowering stems to up 50 cm tall. The bulbs look like small individual cloves of garlic. The leaves are blade-like, about 30–40 cm long and 5 cm wide, with a hairless midrib. The whole plant smells strongly of garlic, filling woodlands with its pungent scent.

The flowerheads are a sphere of flowers with six pure-white petal-like tepals, each about 1 cm across. The fruits are a cluster of three little green spheres.

This is the most distinctive native garlic, although in urban areas the non-native few-flowered leek may be more common. Ramsons was an important pot herb and is still very popular with wild foragers as a salad or in cooking. For traditional medicinal uses, it would be a suitable substitute for the more familiar bulb garlic.

Ramsons were used on Arran in the form of an infusion to combat kidney stones, otherwise known as "the gravel" or "the stone". Martin Martin (1703) describes it as follows:

> *Allium latifolium [A. ursinum],* a kind of wild garlic, is much used by some of the natives, as a remedy against the stone: they boil it in water, and drink the infusion and it expels sand powerfully with great ease.

It was also used widely, as were other onions and garlics chopped finely, then mixed with butter to draw the pus from boils. Mary Beith (2004) mentions it being mixed in with chopped foxglove leaves for this purpose.

This biennial herb has a basal rosette of soft, downy, slightly grey leaves shaped like lance heads. These grow to around 30 cm long, and remain over the winter. In its second year, the plant sends up the characteristic spike of purple-pink thimble-like flowers to between 1 and 1.5 m tall.

Foxglove features in one of our most peculiar magical recipes: a love charm made by burning it along with butterbur (*Petasites hybridus*), seaweeds, royal fern (*Osmunda regalis*), and the bones of an old man. The burning had to be carried out on a flat rock by the seashore, then the ashes were rubbed on the chest of the charm-maker's lover to ensure a long and faithful relationship. This charm was recorded in Alexander Carmichael's *Carmina Gadelica*, accompanied by the following instructions.

Left: Foxgloves by Jacqui Pestell MBE, watercolour (contemporary).

Foxglove *Digitalis purpurea*
Plantaginaceae (formerly Scrophulariaceae)

English:	Bloody bells, Bloody fingers, Dead-men's bells, Fairy's thimbles, King Ellwand's, Lady's thimbles, Scotch mercury, Tod's tails, Wild mercury, Witch's paps
Scots:	Bluidy bells, Bluidy fingers, Fox trie leaves, Foxter, Gensie Pushon, Lady's thummles
Gaelic:	Lus nam Ban-sith. Many names for foxgloves have sinister overtones that imply death, witches, or the faerie folk.
Flowering:	June to August
Habitat:	In woodlands – especially clearings, margins and banks

Arise betimes on Lord's day, to the flat rock on the shore, Take with thee the butterbur and the foxglove. A small quantity of embers, in the skirt of thy kirtle, a special handful of sea-weed In a wooden shovel. Three bones of an old man, newly torn from the grave, Nine stalks of Royal fern, newly trimmed with an axe. Burn them on a fire of faggots, and make them into ashes; sprinkle in the fleshy breast of thy lover, against the venom of the north wind. Go round the rath of procreation, The circuit of five turns, And I will vow and warrant thee that man [woman] shall never leave thee.

Martin Martin (1703) alludes to some painkilling properties: "Foxglove, applied warm plaisterwise to the part affected, removes pains that follow after fevers." Lightfoot (1777) explains that the bitter qualities of this plant made a decoction of the leaves suitable for use as an emetic and cathartic, and an ointment for treating the scrofula, a lymphatic infection also known as 'the king's evil'.

Right: Foxgloves by Jan Kerr, watercolour (contemporary).

Main and right: Preserved specimen of cleavers, RBGE Herbarium, http://data.rbge.org.uk/herb/E00907156.

Cleavers *Galium aparine*
Rubiaceae (madder family)

English:	Cleavers, Goosegrass
Scots:	Bloodtongue, Bleedytongues, Catch-rogue, Catchweed, Grip-grass, Gruppit grass, Guse-grass, Lizzie-rin-the-hedge, Loosy-tramps, Robbie-rin-the-hedge, Stickers, Sticky-grass, Sticky-willie, Sticky-willow
Gaelic:	Garbh-lus
Flowering:	June to August
Habitat:	Hedgerows and woodland margins, especially by disturbed paths and roadsides; common in urban areas

This well-known annual plant scrambles to about 1 m long. The simple leaves grow up to 7 cm long in whorls of six to eight at the nodes. The small white flowers are borne on a cyme and are tubular, with four petals in a cross shape. The flowers eventually produce a fruit that looks like nothing so much as a tiny green bum. The best known feature is the hooks that cover the entire surface of the plant, helping aid its dispersal by catching on to passing animals. The plant is a relative of coffee, and the fruits were used widely in northern Europe as a coffee substitute. Like other members of the rubiaceae, the roots were a source of a red dye, although Lightfoot (1777) suggests it was so scattered that it was not worth collecting enough to make a dye. Its main use was in children's games: sticking it to others' clothing made for hours of fun and it featured in a cruel game called 'bleedy tongues', in which children would rasp the rough leaf down the tongue of a rival.

Ivy *Hedera helix*
Araliaceae (Ivy family)

Scots:	Binwuid, Eevie, Ivery, Ivri
Gaelic:	Gort
Flowering:	September to November
Habitat:	Woods, parks and gardens, clambering up trees and rocks; can form a free-standing shrub if the supporting tree dies

Main: Ivy by Jessica Langford, watercolour (contemporary).

98 Chapter 4: Woodlands

This perennial evergreen vine uses clinging adventitious roots to climb to 20 m or more. The leaves are a distinctive palmate shape with three to five lobes, sometimes shifting to heart-shaped in older parts of the plant, in fuller-sun or before flowering. These are usually matte, dark green, but there are many hundreds of cultivated varieties, and several other non-native species that have escaped, so the variation is enormous. The flower heads are a sphere of greenish yellow flowers with five triangular sepals and tiny petals, with a nectar disc in the centre on the top of the ovary. When fertilised, they produce the characteristic black fruits.

Ivy was used to treat burns, including sunburn, usually as an ointment made with the twigs or leaves with butter as a base, or in a vinegar mix to get rid of corns.

The leaves produced a black dye, and there are records of this being used as a colour restorer for a policeman's uniform. Dark greens and creamy yellows are also possible with various mordants.

In another of Carmichael's translations of Gaelic charms (1900), we have an interesting allusion to a possible use for ivy in animal husbandry:

> I will pluck the tree-entwining ivy, As Mary plucked with her one hand, As the King of life has ordained, To put milk in udder and gland, With speckled fair female calves, As was spoken in the prophecy, On this foundation for a year and a day, Through the bosom of the God of life, and of all the powers.

Presumably, this was an incantation addressed to the ivy as it was being harvested. There is a very widespread concept, almost worldwide, that a charm or prayer should be uttered to the plant being collected or the animal being hunted. In some cases this was done out of respect for the organism, in others to avoid retribution, to maximise its effectiveness, or

Above: Ivy, John Hutton Balfour teaching diagram collection (19th century), showing the climbing roots.

just because it had always been done that way. Unfortunately in this case we do not know the motives behind these verses, nor can we easily understand the mentions of prophecy, or why the ivy should be collected one-handed. Perhaps it does not matter, and we are extremely lucky that such verses and prayers were recorded, but they all too often prompt far more intriguing questions than they answer.

St John's wort
Hypericum perforatum and *H. pulchrum*

Hypericaceae (St John's Wort family)

Gaelic:	Beachnuadh boireann (*Hypericum perforatum*), Lus-Chaluim-Chille (*H. pulchrum*)
Flowering:	July to September
Habitat:	Dappled woodland, clearings and woodland edges as well as acid grassland

Left: St John's wort, John Hutton Balfour teaching diagram collection (19th century).

This perennial herbaceous plant reaches 50 cm high and has opposite leaves and branches. The leaves are simple, about 5 cm long, with untoothed margins, lanceolate in overall shape and rounded at the tip. When the plant is held up to the light, the perforatum of its taxonomic name can be seen – tiny transparent circular windows. The flowerheads are an open dome of stalked flowers, each about 2 cm across, with five golden-yellow petals and numerous stamens forming a fringe in the centre. These surround a central ovary, which forms a dry teardrop-shaped capsule of a fruit.

St John's wort was one of the most important plants in traditional lore in the Gaelic-speaking areas of Scotland, bridging the gaps between religion, magic and medicine. It was widely used throughout Europe, and the details of its use and the context in which it was used were tailored to the local beliefs in particular. Lightfoot (1777) summarises this:

> The superstitious in Scotland carry this plant about them as a charm against the dire effects of witchcraft and enchantment. They also cure, or fancy they cure their ropy milk, which they suppose to be under some malignant influence, by putting this herb into it and milking afresh upon it.

One of its Gaelic names, lus na fala, relates to its use in treating bleeding wounds, and Mary Beith (2004) explains that it was boiled and held against the wound. This plant's even more intriguing name, achlasan Chaluim Chille, is often translated as 'armpit package of St Columba'. Beith explains how St Columba helped cure a shepherd boy of his fear of the dark nights by telling him to put the plant under his armpit. Martin Martin (1703), during his time on Lewis, saw a man who was said to have the second sight. By wearing a piece of St John's wort (possibly a related species, *H. pulchrum*) in his armpit, the apparently distressing condition was stopped.

This species is also named Mary's plant, and in Aberdeenshire, sleeping with the plant under your pillow was believed to grant a vision of St John, who would grant a blessing and protection through the coming year. Alexander Carmichael (1900) gives a fairly extensive account, which draws many of these elements together:

> St John's wort is known by various names, all significant of the position of the plant in the minds of the people. St John's wort is one of the few plants still cherished by the people to ward away second-sight, enchantment, witchcraft, evil eye, and death, and to ensure peace and plenty in the house, increase and prosperity in the fold, and growth and fruition in the field. The plant is secretly secured in the bodices of women and in the vests of men, under the left armpit. St John's wort, however, is effective only when the plant is accidentally found. When this occurs the joy of the finder is great and gratefully expressed: 'St John's wort, St John's wort, Without searching, without seeking! Please God and Christ Jesu This year I shall not die.' It is specially prized when found in the fold of the flocks, auguring peace and prosperity to the herds throughout the year. The person who discovers it says: 'St John's wort, St John's wort, Happy those who have thee, whoso gets thee in the herd's fold, shall never be without kine.' There is a tradition among the people that St Columba carried this plant on his person because of his love and admiration for him who went about preaching Christ, and baptising the converted, clothed in a garment of camel's hair and fed upon locusts and wild honey.

Many of the actions alluded to by the stories may be associated with the antidepressant effect of St John's wort. The mechanism is still unclear, but the effect is well documented.

It was also a useful dye plant, giving yellow dye with alum as a mordant and a red dye from the tops with alum and tin. *Hypericum perforatum* infused in oil will turn the oil a rich red, and this form of extraction is often used by modern herbalists.

102

Wood sorrel *Oxalis acetosella*

Oxalidaceae (wood sorrel family)

English:	Cuckoo's meat, Cuckoo sorrel, Lady cakes, Lady's clover, Poor man's lettuce, Shamrock, Sour clover
Scots:	French sourock, Gowk's meat, Gowk's clover, Sheep soorag, Sookie-sooriks, Sourock
Gaelic:	Feada coille
Flowering:	May to June
Habitat:	Grows in drifts in shady woodlands of all types, typically among moss

This perennial, patch-forming herb has leaves arising from a rhizome up to 10 cm tall. The leaves are trifoliolate, with scattered long hairs. The flowers are solitary, slightly nodding with white and pink-veined petals, and form a capsule in fruit.

The leaves of wood sorrel were employed throughout Scotland, Ireland and other north European countries as a salad vegetable, as well as to make a 'tea'. The content of oxalic acid made the drink sour and, as a consequence, refreshing, so it was thought to be particularly good for heatstroke or fever patients. In large quantities, however, the acid can be toxic, and there are historical records of accidental poisonings in very young children. That said, it was a very popular impromptu snack for children (and still is for those in the know). Alexander Carmichael (1900) listed it with primrose as "The children's food in summer; Geimileachd, geimileachd, Wine and plovers, The food of men in winter."

"Wine and plovers" refers to the most luxurious of winter foods, which primroses and wood sorrel were to children during the summer months.

Lightfoot (1777) mentions that the whole plant was eaten in his day, explaining that it has "an agreeable acid taste and cooling quality". He specifies a more medicinal use against scurvy and, on Arran, against "malignant" and "putrid" fevers.

Wood sorrel's common names revolving around cuckoos, fools and gowks are all intertwined. A fool in Scots is usually 'gowk', which is also a name for the bird cuckoo (*Cuculus canorus*). Plants with cuckoo in the name appear or flower in early May, when the cuckoo arrives from its winter migration. Alternatively, the 'fool' referred to here may derive from the cuckolded bird in whose nest the cuckoo lays its eggs.

Carmichael and several of the common names in Gaelic, Scots and English conflate this plant with *Trifolium*, the clovers that are more commonly considered shamrocks. It may be that they were interchangeable in lessons about the Trinity and St Patrick.

Left: Wood sorrel by Fiona Ward, watercolour (contemporary).

Above and right: Bramble by Jessica Langford, watercolour (contemporary).

Brambles and raspberries
Rubus fruticosus and *R. idaeus*
Rosaceae (the rose family)

Rubus fruticosus	
English:	Blackberries, Black buttons
Scots:	Black boyds, Blackbides, Blackbutters, Brammie, Brammle, Bumblekites, Drumlie-droits, Garten-berries, Gatter-berries, Gatter-tree, Lady garten-berries, Scaldberries
Gaelic:	Dris

Rubus idaeus	
Scots:	Hindberry, Rasp, Siven, Thummles, Wood rasp
Gaelic:	Sùbh-craoibh
Flowering:	May to July
Habitat:	Often in disturbed areas near woodlands, hedgerows and in urban parks and gardens

Chapter 4: Woodlands

These perennial semievergreen shrubs have stems to 1.5 m. These are arching and armed with thorns in bramble and upright canes with prickles in raspberry. The leaves are compound-palmate in bramble and pinnate in raspberry, with leaflets usually ovate to lanceolate and with toothed margins. The rose-like white- or pink-petalled flowers on both species grow to about 4 cm across (usually less), and form the classic bobbly aggregate fruit.

These species are both native to Scotland, and have long been foraged for their delicious berries. We can be fairly certain that Mesolithic hunter-gatherers ate them, a practice that continues with modern foragers. A traditional tale told throughout Britain is that the devil is said to spit or urinate on the berries after Michaelmas (29 September), because he was thrown from Heaven by St Michael on that day.

Bramble and raspberry have also been domesticated and are a major part of the berry economy of the Tay valley, with modern breeding programmes at the James Hutton Institute at Invergowrie concentrating mostly on raspberry and hybrids.

Buchanan White (1876) states that, "The berries are often eaten and afford a good jelly." Withering (1776) notes that, "The berries when ripe are black, and do not eat amiss with wine. The green twigs are of great use in dying woolen, silk and mohair black. Cows and horses eat it; Sheep are not fond of it." They also commonly find their way into historical and modern alcoholic drinks – from wines to gins. Henslow (1905) writes that, "It is said that the inhabitants of Skye use them for making syrup and spirituous beverages." The juice of raspberries on their own was used for a refreshing lightly distilled drink popular following a fever.

From a more medicinal perspective, wild raspberries that were boiled and flavoured with mint and bogbean were given to a patient after a bout of jaundice, and the roots of brambles, infused in water with pennyroyal (*Mentha pulegium*), were used in the treatment of bronchitis and asthma.

Bramble leaves were used as an astringent to treat bacterial infections, and Lightfoot (1777) particularly refers to erysipelas (a group A streptococcal infection). The leaves of raspberry were, and still are, widely used in a tea by expectant women, the belief being that the drink would strengthen the muscles of the womb.

Both species were and are used for dyeing, and Margaret Bennett, the renowned Gaelic folklorist, records bramble switches being hung above door lintels to ward off the evil eye.

Woodland trees

Trees are often the most emotive and fascinating of plants for people. Even the few species native to Scotland are immensely versatile beyond their use as timbers. Hugh Fife's *Lore of Highland Trees* (1985) is difficult to come by, but is an excellent account of their uses. Many other writers have described woodlands and woodland practices in Scotland.

Almost all of our 20 or so native tree species were important to the domestic economy. Some are lumped together as generic groups, with the two tree birches often treated as one species in terms of use, and many of the shrubby or tree willows share similar properties. Oddly, perhaps the least 'important' trees are the distinctive holly (*Ilex aquifolium*), and elm (*Ulmus glabra* and related species). Holly makes a useful forage for animals, and the wood has a fine grain, so was often used for making printing blocks or small items as such as boxes and fine furniture, but it perhaps has fewer quirky and local uses than other species. That said, it is a very distinctive plant, so the Gaelic name chuilinn appears in many place names, perhaps as a reference to traditional landmark trees or stands of holly. A humorous phrase recorded in Roy Vickery's *A Dictionary of Plant Lore* (1997) suggests that an untrustworthy character "never lees (lies) but when the holyn's green". Elm wood was durable and used for pipes, wheels and general construction.

The trees in this chapter are mainly from a typical broadleaved woodlands, although in the central belt these are now often dominated by non-native sycamore (*Acer pseudoplatanus*). Other species are listed under more appropriate habitats, for example Scots pine (*Pinus sylvestris*) and rowan (*Sorbus aucuparia*) are in the 'Moorland and Mountains' chapter.

Left: Holly from Sowerby's *English Botany*, Vol. 7 (1798).
Right: Pendunculate oak by William Phillips, watercolour (contemporary).

Chapter 4: Woodlands

Gaelic alphabet of tree names

Ailm (elm) = A
Beith (birch) = B
Coll (hazel) = C
Dair (oak) = D
Eadha (aspen) = E
Fearn (alder) = F
Gort (ivy) = G
(H)uath (hawthorn) = H
Iogh (yew) = I
Luis (rowan) = L
Muin (vine) = M
(sometimes interpreted as bramble or honeysuckle)
Nuin (ash) = N
Oir or Onn (gorse/whin) = O
Peith bhog (downy birch) = P*
Ruis (elder) = R
Suil (willow) = S
Teine (whin/gorse) = T (based on an archaic letter)
Ur (heather) = U

*This is a more recent adaptation than the others.

In a particularly interesting cultural use, the Gaelic language matches the names of the letters of the alphabet to names of trees and shrubs, creating the much-celebrated *Gaelic alphabet of tree names*. Its origins are uncertain, but it is thought to have been a teaching aid. The system can often be seen interpreted for visitors to woodlands throughout the country.

Silver birch *Betula pendula* and **downy birch** *B. pubescens*

Betulaceae (the birch family)

Scots:	Birk, Birken tree
Gaelic:	Beatha
	B. pendula: Silver birch, Black birch, Knotty birch, Beith dubach
	B. pubescens: Downy birch, Beith charraigeach and Beithe dhubh-chasach
	Betula nana: the small montane shrub Beatha beag
Flowering:	April
Habitat:	Throughout Scotland to relatively high altitudes. The silver birch is more typical of drier sites, whereas the downy birch is from wetter 'carr' habitats.

This tree grows to 25 m. The bark is silvery and smooth when young, breaking into dark diamonds when older. The downy young twigs give *B. pubescens* its name. Those on *B. pendula* are warty, without hairs. The simple and triangular leaves are spirally arranged on the branch, somewhat rounder and with single teeth in *B. pubescens*; they have double teeth in *B. pendula*. Both species have green pendulous catkins in early spring.

This is undoubtedly one of the most long-used and versatile native plants in Scotland. Stakes made of birch and sharpened with stone axes have been found placed upright in the bottom of pits (thought to be pitfall traps) in the Wigtownshire area. These have been dated to around 2,500 BCE. The wood was also used for making the floors of crannog loch-houses, but it was held above the water because it is not particularly resistant to rotting.

In more recent times, the wood was used for fish barrels, wheels, carts, ploughs and, as Lightfoot (1777) puts it, "most of the rustic instruments". Turners made bowls and other items from the wood, which was also burned as fuel and for charcoal. Smaller branches were used for fencing, and switches as brooms for "the purposes of cleanliness and correction". During the boom times of the textile industries in Scotland in the industrial revolution, bobbins for spinning were made from birch. Demand was so great that the birch used was both home-grown and imported from Canada and Scandinavia. The 'outer rind' (Gaelic: meilleag) was burned as a substitute for candles, and young whippy twigs could be twisted together into strong ropes, an excellent example of which can be seen in the National Museums of Scotland.

Above and left: Birch illustration by Sarah Howard, watercolour (contemporary).

In the culinary world, herring and hams were often smoked using birch branches, and the wood was used to keep the fires going when distilling whisky.

Thomas Pennant (1774), who travelled with Lightfoot, remarks that "A great deal of excellent wine is extracted from the live tree." Although there is perhaps a miswording here, because the plant does not ferment the sugars itself, it is certain that birch-sap, or 'birk' wine was a popular beverage in Scotland. Johnson, writing in 1862 has this to say on taking sap from birch:

> From a flourishing tree of moderate size from four to six quarts may be obtained in a single day… The fresh sap as it is extracted from the tree forms a very pleasant drink, and is supposed by the Highlanders to be very beneficial in complaints of the bladder and kidneys.

McNeill's *The Scots Kitchen* (1929) gives this recipe for birch wine:

> To every gallon of the juice from the birch tree, three pounds of sugar, one pound of raisins, half an ounce of crude tartar, and one ounce of almonds are allowed; the juice, sugar and raisins are to be boiled twenty minutes and then put into a tub, together with the tartar; and when it has fermented some days, it is to be strained and put into the cask, and also the almonds, which must be tied in a muslin bag. The fermentation having ceased, the almonds are to be withdrawn, and the cask bunged up, to stand for about five months, when it may be fined and bottled. Keep in a cool cellar. Set the bottles upright or they will fly.

Below and right: Hazel by Geoffrey Brown, watercolour (contemporary).

110 Chapter 4: Woodlands

This small tree, or more commonly multistemmed shrub, has simple, broad, round, alternate, hairy leaves with toothed margins. The young twigs have bristly red hairs and the bark is silvery and smoothish when young, with large lenticels. The male catkins are pendulous, soft and yellow. The female catkins are like a small bud with tentacle-like stigmas protruding; these form the classic hazelnuts over the course of the year.

Coppicing (cutting the stem back to produce vigorous, pliant shoots) hazel produced the ideal rods for wattles and hurdles as fences, lathwork for plaster, spars in thatching, and the shanks for walking sticks, crooks and wading sticks.

Martin Martin (1703) gives an intriguing account of the use of hazel in a sweat treatment akin to the sweat-lodges of the Scythians or Native Americans. For more information see under Soft rush in the 'Wetlands' chapter.

Hazel had a good reputation in occult, magical and religious contexts. A hazelnut was a symbol of knowledge, the salmon of knowledge in the Irish Fenian cycle having eaten nine hazelnuts before passing the knowledge on to humanity. St Kentigern used a hazel branch to rekindle a fire that had gone out, and twin hazelnuts were thought to bring good luck.

Hazel *Corylus avellana*
Betulaceae (the birch family)

English:	Crack-nut, Cracker nut, St. John's nut
Scots:	Hasil, Haslie, Hassel, Hassil, Hasslie, Hassly, Hazellie, Hissel, Scob, Scub
Gaelic:	Calltainn
Flowering:	December to March
Habitat:	Throughout all woodlands

Through the early and mid 20th century, hazelnuts were still collected by the young to help them divine the future. Hazel has also long been used for another form of divination, as explained by Hooker in *Flora Scotica* (1821):

> It is of the young forked twigs of this plant that the celebrated divining rod (virgula divinatoria) is taken, with which individuals even in our days and our country have believed that they possessed the power of discovering springs of water, when nothing on the surface of the earth indicated their existence.

Hawthorn *Crataegus monogyna*

Rosaceae (the rose family)

English:	May
Scots:	Boojuns, Chaw, Cheese-an-breid, Fleerish, Flourish, Has tree, Hathorn, Haw, Haw-berry, Haw-bush, Leddy's meat
Gaelic:	Sgitheach
Habitat:	Hawthorn is widespread throughout the country, in hedgerows, woodlands, parks and gardens from sea level to around 500 m.
Flowering:	April to June, but mostly May, hence the common name.

Left and right: Hawthorn by Nichola McCourty, watercolour (contemporary).

This small deciduous tree or shrub grows to 10 m. The bark is brown and grey and platy, flaking off with age in either layers or in dark reddish strips. The leaves are spirally arranged on short shoots, which can be spiny. The flowers are around 1.5 cm across, with white petals in dense domed heads. The fruits are the distinctive dark-red 'haws'.

Hawthorn is perhaps best known as a hedging tree, and is a major component of hedgerows. As Hooker's *Flora Scotica* (1821) explains, "It is excellent for fences, and bears clipping admirably."

The wood was also valued for making small, decorative items, household tools and handles for larger implements.

Mary Beith (2004) mentions its use medicinally for sore throats (as a decoction) and for treating imbalance in blood pressure.

The young leaves were, and are, eaten as a snack, hence some of the names relating to cheese and meat. Although the haws are unpalatable raw, they make an excellent jelly.

The bark yields a black dye with copperas (iron sulphate) and, according to modern dyers, the leaves give an olive brown (with alum as a mordant).

As a divinatory aid, the flowering of hawthorn was a guide on when to put away the winter clothes. As the rhyme advises, "ne'er cast a cloot 'til May is oot".

Like several other plants, it was believed to be unlucky to have hawthorn in the house, or even the garden because the flowers allegedly smell of death. It appears often as trysting tree in tales of faerie lore. Indeed, 'clootie trees' (trees hung with cloths) were often hawthorns.

Above and right: Ash by Kathy Munro, watercolour (contemporary).

Ash *Fraxinus excelsior*
Oleaceae (the olive family)

Scots:	Esch, Esh
Gaelic:	Uinnseann
Flowering:	May
Habitat:	Lowland broadleaved woodlands, more common in the East

114 Chapter 4: Woodlands

This tree grows to 35 m. The bark is silvery and smoothish when young, breaking into deep fissures in older plants. The leaves are opposite, compound and up to 30 cm, with between three and six pairs of dark-green leaflets tipped with another leaflet. The flowers are tiny but grow in prolific heads on separate male and female plants. The fruits on the female trees become the characteristic and easily recognisable winged 'keys'.

The wood of ash is strong, but springy, and appears in epic poetry from the *Iliad* to the *Mabinogion* as the shafts of spears. Throughout Scotland it was used for the same job, as well as being essential for agricultural tools, tent pegs or anything that would take a thump. It turns well for bowls and chair legs.

Ash was tapped in a similar way to birch in the past, and the sap of a young ash twig was part of a ritual for baby boys. A green twig would be held in the fire until the sap bubbled out, then the sap was dropped in the baby's mouth.

There is a suggestion that this may be a remnant of the Norse influence on Highland tradition. Yggdrassil (the 'steed of the terrible one') was the World Tree of Norse mythology – the fundamental organism holding the worlds of gods, men and giants together. Odin himself hung from the World Ash to learn the magic of runes. By giving the sap to a newborn baby, it may have been believed that some of the mystical power would be transferred to the child.

"Theid an nathair troimhn teine dhearg mun teid i troimh dhuilleach an uinnsinn." Mary Beith (2004) records this charm, which means, 'The snake will go through the red blazing fire rather than through the leaves of the ash.' This may relate to the Norse myth of Niddhog, the dragon at the base of the Norse World Ash.

"Ash before oak, the lady wears a cloak, Oak before ash, the lady wears a sash." This English rhyme uses ash as a divinatory tool to predict the upcoming summer weather, depending on which of two trees comes into leaf first.

Chapter 4: Woodlands 115

116 Chapter 4: Woodlands

Cherries and sloes
Prunus species

Rosaceae (the rose family)

Scots:	Hackberry, Hagberry (*P. padus*), Chirry, Gean, Guind, Merry-tree, Sirist (*P. avium*); Bulister, Slae (*P. spinosa*)
Gaelic:	Geanais, Fiodhag (*P. padus*), Preas nan airneag (*P. spinosa*)
Flowering:	May, June; sloe flowers appear in April
Habitat:	Lowland broadleaved woodlands and hedgerows

Main: Blackthorn by Sarah Howard, watercolour (contemporary).

This is a fast-growing deciduous group of spreading trees. *P. padus* and *P. avium* grow to around 15 m tall, and have distinctive dark brown bark with large orange, horizontal lenticels. The leaves are spirally arranged and simple, about 15 cm long, and are toothed, often flushed with red, and hairless. The flowers are like small roses. In *P. avium* they grow to 2 cm across and in a cluster, but are smaller and on racemes in *P. padus*. The fruits are cherries and larger in *P. avium* than in *P. padus*. *P. spinosa* (sloe) is a smaller shrub with small leaves and waxy-bloomed fruits.

The wood of cherry is an attractive orange or red. It was often used for making small boxes, turned items or detailing, although it was considered unlucky to fell cherries – and gean in particular. Bird cherry wood was not used for staves or other purposes in the north-east, because it was said to be a witch's tree. Similarly, blackthorn was associated with the shidhe and other evil spirits. Children stolen by the faerie folk were left under these shrubs and would grow up to become changelings. Pricking oneself on a sloe thorn could bestow a curse.

The fruits of all these species have long been used for flavouring spirits. Lightfoot (1777) names the tart fruits of bird cherry (*P. padus*) as particularly popular in brandy, and Hooker (1821) says of blackthorn that the fruit is, "small, very austere. Used to adulterate Port wine, as the leaves are to mix tea."

'Gum', which may be a bacterial exudate, of the gean was used for treating colds when dissolved in wine. Sloe jelly was used for throat problems, and the flowers of sloe were used as a laxative or infused then applied to the skin to kill off scabies.

Blackthorn bark was used to produce a bright red dye. With the addition of vitriol or copperas, the 'juice' was said to make good ink or blue and black dyes.

Oaks *Quercus* species

Fagaceae (the oak family)

Scots:	Aik, Ak, Eak, Moss aik, Puggie pipe, Knappers (the galls)
Gaelic:	Darach, pedunculate oak is Darach Gasagach
Habitat:	Widely planted, but native through diverse ancient broadleaved lowland woodland, and the beautiful Atlantic oakwoods of the west coast

Main and right: Pedunculate oak by Jan Kerr, watercolour (contemporary).

118 Chapter 4: Woodlands

These deciduous trees grow to a maximum of 35 m. The bark is grey and fissured. The leaves are spiral (alternate) and simple but with distinctive wavy margins. The flowers grow in long catkins; the male resembles a tiny typical flower, and the female is tucked into a bud made of scales, which become the scales on the cupule of the acorn.

Oakwoods are iconic, diverse and fascinating habitats. The Atlantic oakwoods in particular tend to be sessile oak (*Q. petraea*), with the larger pedunculate oak (*Q. robur*) in the south and east. This latter species has larger acorns and was extensively planted during the 18th century. Much of the oakwoods were managed throughout this time, following a pattern called 'coppice with standards', which ensured there was enough bark for tanning. Woodlands were split into a 24-year rotation system in which each year most of the trees in one of the 24 lots were cut down to a stool, which would then resprout to produce 'tanbark coppice'. The 'standard' part of the name refers to some of the trees that were left to grow without coppicing. Bark would be stripped in May and June (it was said to be 'running'). Large quantities of sessile oak bark were sent from the west coast to Glasgow for the tanning industry. The wood was converted to charcoal and used extensively in iron smelting.

Martin Martin (1703) writes about oak's use in producing yeast:

YEAST, HOW PRESERVED BY THE NATIVES

> A rod of oak, of four, five, six or eight inches about, twisted like a with, boiled in wort, well dried and kept in a little bundle of barley straw, and being steeped again in wort, causeth it to ferment, and procures yeast: the rod is cut before the middle of May, and is frequently used to furnish yeast; and being preserved and used in this manner, it serves for many years together. I have seen the experiment tried and was shown a piece of a thick with, which hath been preserved for making ale with for about twenty or thirty years.

The method was confirmed by James Robertson some seven years later.

Lightfoot explains that oak was called the 'king of the forest' by the Highlanders. The bark was used for a brown or, with iron sulphate, black dye. A gargling solution made from the powdered bark of the oak was used for treating sore throats.

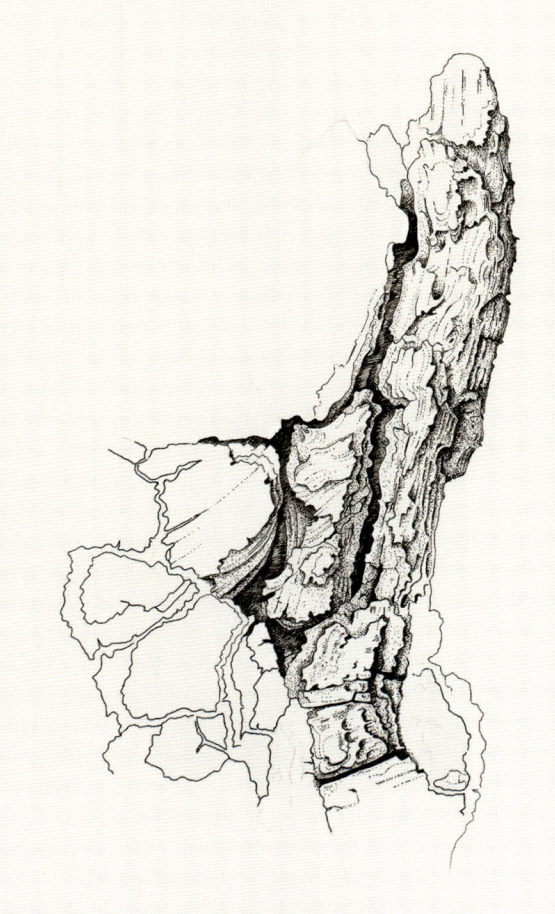

Chapter 4: Woodlands 119

Elder *Sambucus nigra*

Adoxaceae (the elder family, although taxonomy remains unstable)

Scots:	Boon-tree, Bore tree, Borral, Bountree, Bourtree, Eller
Gaelic:	Droman
Flowering:	May to July
Habitat:	Broadleaved woodland, hedgerows and in urban areas including parks, gardens and by railways

This shrub grows to a maximum of 7 m. The bark is corky and deeply fissured when young, breaking into scaly plates when older. The leaves are opposite and compound, with between five and nine leaflets, which are broadly lanceolate with teeth on the margins. The flowers grow in a flat-topped cyme of small (less than 1 cm) whitish flowers with distinctive pop-up stames. The fruits are the characteristic elder berries.

Several of the common names for elder refer to the pith of the wood, which is soft and can be poked out to make versatile pipes. These were used as practice whistles, for taps (e.g., for tapping birch sap or casks), and as stems of pipes for smoking. In larger trunks there is plenty of wood, which was used for small utensils such as spoons, and the traveller families throughout Scotland made excellent examples of these.

Left: Field illustration of elder by Jane Wisely, watercolour (1974).

Elder cordial and elderberry wine have been recorded since the 18th century at the very least and remain popular today. Buchanan White (1876) wrote about the culinary uses of elder thus:

> the cluster of flower buds is said to make a delicious pickle to eat with mutton. Tea, even (which cannot, however, be recommended), has been made from the dried flowers. It is said not to be prudent to sleep under the shade of the tree, from its narcotic properties.

Like rowan (*Sorbus aucuparia*), elder was said to keep witches at bay, so it was planted outside houses in much the same way. The sap from under the bark was said to activate the second sight in those who were prone to it, and standing under an elder tree on a faerie knowe on Halloween could apparently allow a seer to see the faerie party trooping past.

The slightly foetid smell of the flowers was reputed to keep flies away, so pieces were often kept around dairies or worn when out walking. F. Marian McNeill's *Silver Bough* (1957) mentions hillwalkers born in the Highlands putting a sprig in their buttonholes because "Flies and things don't like it."

Lightfoot (1777) lists a number of medicinal uses: an infusion of the inner bark in white wine as a gentle cathartic and deobstruent to get bodily fluids moving again; the leaves, bruised then used as a cataplasm in combating pleurisy; the dried flowers as a sudorific; and the juice of the dried berries as an aid against indigestion (by inducing flatulence).

Right: Preserved specimen of elder, RBGE Herbarium, http://data.rbge.org.uk/herb/E00894515.

Yew *Taxus baccata*
Taxaceae (the yew family)

Scots:	Snot-globs (the fleshy arils)
Gaelic:	Iubhar
Flowering:	Cones in April and May; female 'berries' ripen over summer
Habitat:	Woodlands and old plantings

Above and right:
Preserved specimen of Yew, RBGE Herbarium, http://data.rbge.org.uk/herb/E00399750.

The plant is evergreen, forming a medium-sized but dense tree, often with many trunks as it ages. The bark is shaggy red-brown. The needles are dark, soft and blunt, around 2 cm long. The cones appear on the separate male and female trees. The male cones look like millions of tiny yellow 'cauliflowers', producing billions of pollen grains, and the female cones begin as tiny stubs, which develop into single-seeded cones, surrounded by a red, fleshy aril (an extra seed covering).

Yew is one of our few conifers. We are not sure if it dispersed here naturally or was brought by early humans, but it has certainly been in Scotland since pre-Roman times because there are ancient trees still alive from then. Nowadays, it is almost always found near to human habitation and is especially associated with churches and woodlands that have seen some degree of human management. There are some sites where the management is so ancient that the woods appear natural.

Yew wood has a lovely marbled grain of pale yellow and deep orange laced with black, but it is famously poisonous, and there are several modern stories of wood turners feeling ill after working with it. The only edible part of the tree is the red, fleshy aril around the outside of the seed. This aril has a very glutinous flesh, giving it the unappetising name of 'snot-globs'. The seeds themselves, the foliage and the wood are all toxic, so it is best to just not eat any part of it.

One of the Gaelic names for yew is 'Ioghar', thought by Cameron (writing in 1883) to stem from the fact that the sap was widely used as a poison to tip arrows (iogh = a severe pain). Certainly, the Latin name 'taxus' is allied to the word 'toxin' and Greek term for an archer is 'toxes'. Of course, the use of the springy, strong wood in bow-making is one of yew's most famous properties. One of Scotland's earliest artefacts – the Rotten Bottom bow found at the Carrifran Wildwood project in the Scottish Borders – is made of yew, and dates to around 6,000 years ago. Robert Bruce used the trees at Ardchattan priory to make longbows, and indeed, this is one of the many reasons that people associate yews with churchyards. The idea was that on sacred ground this valuable resource would be considered untouchable, except in direst need. Other interpretations are that the yews protected the dead – often interpreted as a throwback to pre-Christian times. Fife (1994) explains that yew lends its name to Iona:

> [It] was almost certainly connected with pre-Christian cult of the yew as observed on the island. The yew cult, as a form of druidism, was absorbed into the growth of Christianity on Iona in the fifth or sixth century.

This blending of pre-Christian and Christian belief is invariably impossible to untangle, but provides endless speculation about the ancient Fortingall Yew in the old churchyard in Glen Lyon, under which Pontius Pilate was said to have played as a child. It is unlikely the Romans would have had a presence so far north so soon after their first sortie into southern Britain, but a wonderful mystique has built up about this very special plant. It is believed to be one of the oldest plants in Europe, and was likely very significant to the local inhabitants. However, it prompts the 'chicken and egg question': was the yew planted on an existing sacred site, or did it become a sacred site, and ultimately a churchyard, because of the yew? This same question could be asked about yews in ancient churchyards throughout their range and we're unlikely ever to have an answer. To add to its magic, this ancient male tree has recently started to develop female cones.

Chapter 4: Woodlands

Below: Golden shield fern (*Dryopteris affinis*) by Marianne Hazlewood, watercolour (contemporary).

Right: Broad buckler fern (*Dryopteris dilatata*) by Jocelyn Anne Rabbitts, graphite on paper (contemporary).

124 Chapter 4: Woodlands

Ferns

Many ferns can be difficult to identify or are rare, but distinctive and common species were often used, and for a surprising range of uses from pragmatic to magical. *Ophioglossum vulgatum*, the adder's tongue fern, is discussed in Chapter 3, Grasslands, and others are mentioned in Chapter 6, Human Habitats, but most species are from woodland habitats and woodland margins. Hart's tongue, which is also mentioned in Chapter 6, is discussed as a parasite treatment by the botanist James Robertson in the 1770s. Quite what this parasite was is uncertain.

> The Revd. Mr Robertson Minister & several inhabitants of Loch Broom informed me that *Asplenium scolopendrium*/Hart's tongue fried with goat's butter was applied by the people as a cataplasm to extract an animalcule which nestling in their legs or other places, produces exquisite pain.

During the Victorian era, ferns were extremely popular as ornamentals, with alpine and small woodland species grossly over-collected for display or to add to personal pressed-plant collections, often threatening wild populations with extinction. For display, the plants were typically kept in 'Wardian' cases (essentially a terrarium for ferns) and in devoted fern houses. In 1996, a wonderful Victorian fernery at Ascog Hall on the Isle of Bute was reopened to the public after a 20-year restoration, and Benmore Botanic Garden's beautifully restored and updated fernery is not to be missed.

Perhaps because of the difficulty in identifying some ferns, we do find references to 'ferns' lumped together as one. For example, Martin Martin (1703), a competent botanist whose writing does distinguish between some species of ferns, mentions it as follows.

Chapter 4: Woodlands 125

Eyes that are blood-shot or become blind for some days are cured here by applying some blades of the plant fern, and the yellow is by [the users] reckoned the best; this they mix with the white of an egg, and lay it on some coarse flax and the egg next to the face and brows, the patient is ordered to lie on his back.

The dedication services for Simprim Parish Church in 1247 were said to include a prayer by David Le Berham, then Bishop of St Andrews, asking for, among other things "the Blessing of Salt, Ashes and Water, and for the mixture of all three". The president of the Berwickshire Naturalists Club interprets this as a plea to ensure that the salt-making and fern burning industries had success in the nearby area. Indeed, 'fern' ash was an important commodity, used as a potash-rich fertiliser, for cloth-bleaching and for soap-making. By the 17th century, much of this had been replaced by the burgeoning 'kelp' industry, substituting bulky brown seaweeds for ferns. The large, common species of *Dryopteris* (the buckler ferns), *Athyrium* and *Polystichum* (shield ferns) were all treated broadly as 'fern' in this way, but perhaps the main fern for ash production was bracken (*Pteridium aquilinum*). It was a major industry, and in 1634, one Patrick Mauld of Panmure was granted a 31-year patent for "the sole and full licence to make and to cause to be made … soap for washing of clothes". This was exclusive to Mauld, and the King also granted him sole licence "to make potases … Of all sorts of ferns and other vegetable things whatsoever, fit for the purpose." There is a record of its having been used as a payment in kind by tenants to a landlord. In Knapdale, Argyll, a "Decreet of Sale of the Lands of Kilmorie … in favour of Sir Archibald Campbell, 1776," mentions that the tenant farmers were required to pay some of the rent in kind, in the form of "sixteen cartloads of pulled fern". The ferns were delivered to the Laird's mansion at Fernoch for use in the building itself and its outhouses.

Horsetails are an interesting group of ferns that look utterly unlike the typical fern. Lacking fronds, they are essentially only branching stems with spore-bearing cones on the tips. They were used for dyeing, mainly yellows and greens, and the silica granules throughout their tissues made them excellent scourers and polishers known as Dutch rushes. Mary Beith (2004) mentions them being simmered and used to wash wounds.

Left: Hart's tongue (*Asplenium scolopendrium*) by Fiona Ward, watercolour (contemporary). This common fern was used in a medicinal beer discussed in the 'Human Habitats' chapter.

126 Chapter 4: Woodlands

Bracken *Pteridium aquilinum*
Dennstaedtiaceae (Bracken family)

Scots:	Brachan, Braikin, Ern fern, Rannoch, Shady bracken
Gaelic:	An raineach mhòr

Although we demonise this species nowadays as invasive, toxic to livestock and with spores suspected to be carcinogenic, bracken was a fundamental part of the domestic economy. Bracken fronds were used throughout Europe as a packing material for delicate items, until certainly the early 20th century. In Scotland, the potato crop was often packed in bracken for shipping.

The burnt ashes of bracken were said, by Lightfoot (1777), to be excellent as a manure for potatoes and were widely used by the crofters of his time. The fronds were also used as frost protection for potatoes, covering the crop when it was laid down for winter. Lightfoot records it as a readily available source of bedding for animals and also for use in beds for humans; it was thought to have insect-repellent properties. Beyond this, it burned very hot, giving a fire well suited to brewing and baking. Frond stems were sometimes used as a thatch, although these were not as durable as other more typical materials such as heather or even straw.

The spreading fronds have been likened to the wings of an eagle, giving the plant its common name of Ern (Scots for eagle) fern. The pattern of the veins in the cut stems have also been variously seen as a double-headed eagle, an oak tree and the initials J.C., standing for Jesus Christ.

The rhizomes are toxic and were thought to cause a disease called the 'trembles' in sheep that ate them.

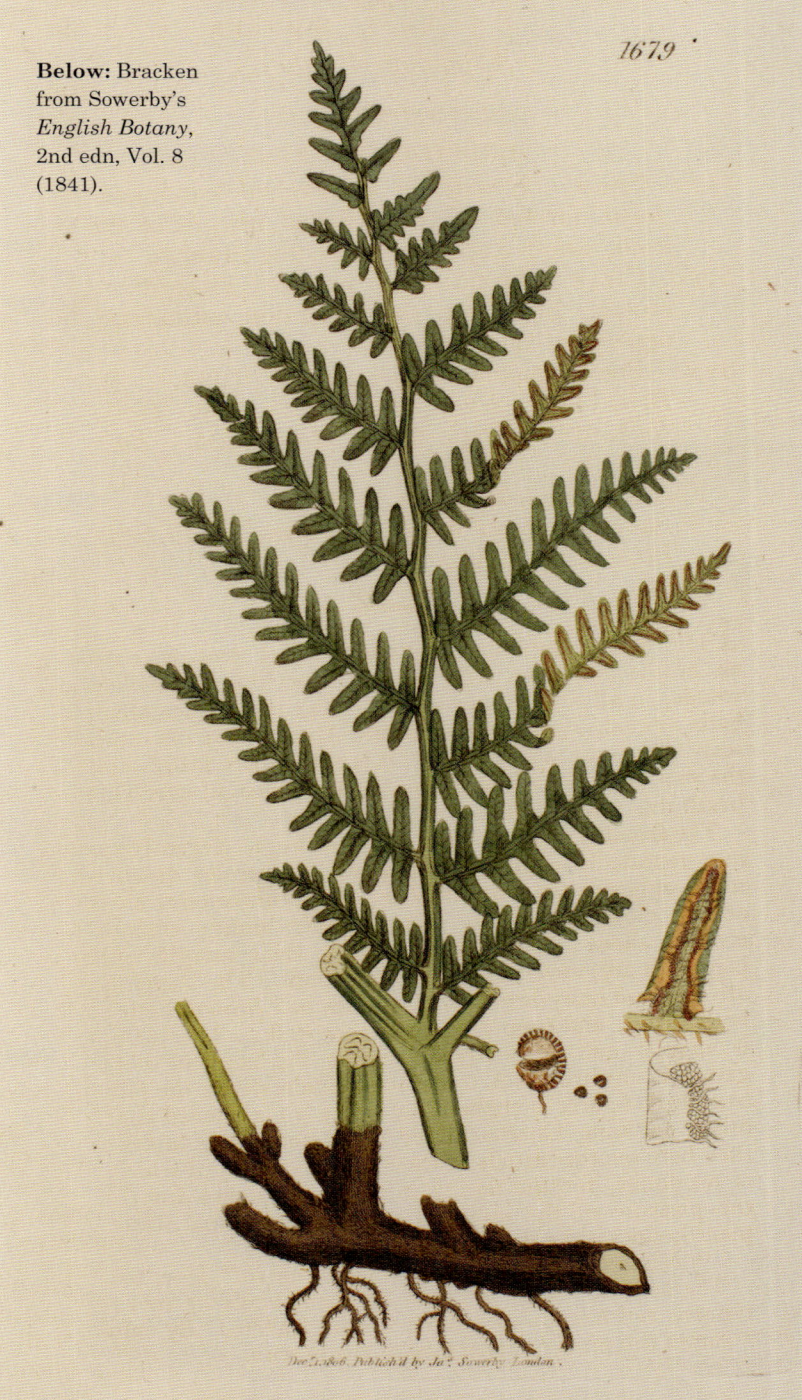

Below: Bracken from Sowerby's *English Botany*, 2nd edn, Vol. 8 (1841).

Chapter 4: Woodlands

Above: Tree lungwort (*Lobaria pulmonaria*) by Claire Dalby RWS RE, watercolour (contemporary).

Lichens

Lichens are composite organisms resulting from the incredible symbiosis between a fungal partner and an alga, with the alga photosynthesing sugars that help sustain the protective fungal layer. Scotland is extremely rich in lichen species, and a long history of research has enhanced our knowledge of these fascinating dual organisms. Traditionally, they found their way into some medicines, but their main use was undoubtedly in dyeing. *Lobaria pulmonaria* (lungwort) was one of the medicinal species. Cameron, in his 1883 *Gaelic Names of Plants, Scottish and Irish,* states that, "It was used among Celtic tribes as a cure for lung diseases, and is still used by Highland old women in their ointments."

Peltigera, the dog lichens, are common on heaths and grasslands, forming a low, often dark mat. They have pointed root-like rhizines on the underside that are suggestive of canine teeth, hence the common name. They were thought by some in the 18th century to be a cure for rabies, perhaps based on the Doctrine of Signatures. In the words of John Lightfoot (1777),

> The L. caninus has a disagreeable musty taste. Half an ounce of the leaves, dry'd and pulverised, and mixed with two drachms of powdered black pepper, compose the once-celebrated Pulvis antilyssus, formerly much recommended by the great Dr. Mead, for the cure of canine madness. This medicine was to be divided in four equal parts, one of which was to be taken by the patient every morning, fasting for four mornings successively, in half a pint of warm cow's milk; after which he was to use the cold bath every morning for a month. It is much to be lamented that the success of this medicine has not always answered the expectation. There are instances where the application has not prevented hydrophobia; and it even uncertain whether it has been at all instrumental in keeping off that disorder.

The lichens known as cudbear on the mainland and korkalit in the Shetland archipelago yield an orchil dye that requires maceration in ammonia solution (traditionally urine) in order for it to work. The dyes that result from these 'orchil acids' are typically reds and purples, but they are often 'fugitive', in that they are not really light- or washing-fast. Lightfoot (1777) records the use, and it was well known through the Harris tweed industry, with boys' urine famously being the best source of ammonia. Cameron, writing in 1883, describes the process:

> This lichen was extensively used to dye purple and crimson. It is first dried in the sun, then pulverised and steeped, commonly in urine, and the vessel made air-tight. In this state it is suffered to remain for three weeks, when it is fit to be boiled in the yarn which it is to colour. In many Highland districts many of the peasants get their living by scraping off this lichen with an iron hoop, and sending it to the Glasgow market.

Modern dyers typically use a commercial source of ammonia, and can recover a red-purple dye if soda is added to the dyebath and a blue-purple if vinegar is added. This series of colour reactions with alkaline or acid was exploited in making the original litmus papers, and an interesting letter in the archive of the Royal Botanic Gardens, Kew suggests it was reconsidered as a source of homegrown dyes during World War II:

> There was formerly a considerable industry in Scotland, gathering the Ochrolechia, largely for use as a dye plant. This is still employed to some extent, and no doubt it would be possible to organize a collection of supplies in Scotland. This would restart a former rural industry.

The other main dyeing lichens belong to the crottle group, which includes several species from the genus *Parmelia*. Again, Lightfoot (1777) explains that it transcended its use as a dye:

> so much did the Highlanders believe in the virtue of crotal that, when they were to start on a journey, they sprinkled it on their hose, as they thought it saved their feet from getting inflamed during the journey.

Doreen McIntyre's excellent thesis on dyestuffs in Scotland (1999) explains that this may be an association between the lichen and the earth: crottle-dyed socks were worn if going on a long journey on foot. However, if sailors wore crottle it could bring bad luck and if they drowned their body would never be recovered. She includes quotes that "anyone wearing crottle dyed garment sinks like a stone" and "What comes from the rocks returns to the rocks."

Known as scrottyie in Shetland, it was collected using a shell as a scraper, or a spoon.

Fungi

Fungi are a kingdom unto themselves, more closely related to animals than to plants. Despite this, their sedentary lifestyle and the way they grow – with what look to be roots under the ground but are actually a mat of fungal hyphae – means they are often lumped in with plants. They have earned a particular air of danger through the relatively few poisonous species, such that Robert Sibbald's *Provision for the Poor in Time of Dearth and Scarcity* does not mention a single edible mushroom

Lightfoot (1777) mentions the cultivation of the field mushroom (*Agaricus campestris*) in late-18th-century Scotland. He also emphasises the toxicity of other mushroom species. Hogg and Johnson (1866), remark on the abundance of chanterelles (*Cantharellus cibarius*) in Perthshire.

We have some interesting accounts of fungal hyphae being used in Ireland to staunch the flow of blood, and Scots in the 1930s used the mould from the top of jam for similar purposes, potentially employing some of the antimicrobial compounds in fungi that became better known as antibiotics.

'Touchwood' (tinder) was commonly made from *Fomes fomentarius* (probably throughout its European range) by paring off the upper 'rind' and boiling the remainder of the fungus in lye. The mass was then dried and pounded with a hammer. Alternatively, it was boiled in saltpetre. This species of bracket fungus was beaten into soft, square pieces and used by surgeons to staunch the flow of blood from minor wounds. This was sold under the name of 'Agaricus'.

Lightfoot (1777) writes that puffballs (various species of *Lycoperdon* and their allies) were sometimes pressed and dried in an oven and then set alight to smoke bees into a torpor in their hives. A similar recorded use suggests the smoke from dried puffballs was also used as a type of anaesthetic for humans. These practices were probably used throughout much of Europe, and there are similar widespread mentions of their use as styptics for staunching the bleeding from wounds.

The offensively named Jew's ear fungus (*Auricularia auricula-judae*) grows on elder trees, and looks like a human ear when hydrated, right down to the texture and downy 'hair' on the surface – eerie indeed. That English common name is thought to refer to Judas Iscariot, as elder is one of many tree species that he is said to have hanged himself from. It is edible if boiled for a long time, and allied species are common in East Asian cooking. It may appear as the 'Black Luggie' mentioned in a poem to keep witches away (see under Rowan in the 'Moorland and Mountains' chapter). An infusion or decoction of this fungus in milk or vinegar was used as a gargling solution in treating 'quincies' (tonsillitis) and sore throats. By Lightfoot's time, the practice was said to be dying out. The ability for this fungus to absorb large amounts of fluids when dry led to its employment as a swab for applying eyewashes.

Left: Fly agaric (*Amanita muscaria*) by Nichola McCourty, watercolour and graphite (contemporary).

Right: Shaggy ink-cap (*Coprinus comatus*) by Fran Thomas, auto-digested ink-cap fungus on paper (contemporary).

Chapter 4: Woodlands

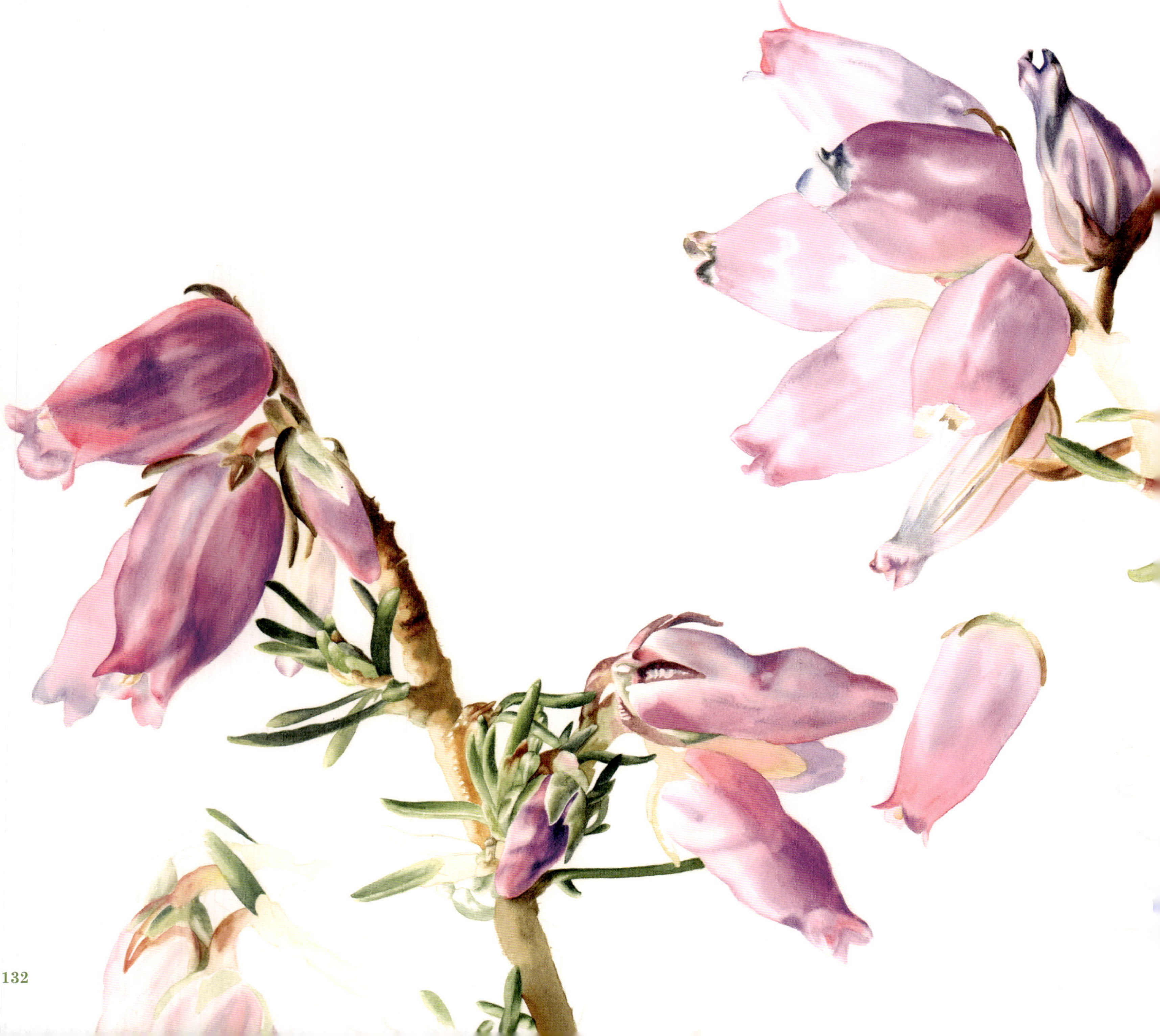

Chapter 5
Moorland and Mountains

Heather moorland covers a huge area of upland Scotland. It is a secondary community, held in place by burning, grazing and other human management. When fenced off and left alone, it is quickly replaced by other shrubs and trees, often reverting to Scots pine woodland in the centre and north of the country. Where they are found together, heather moorland blends into Scots pine forest on gravelly soils, or into acid grassland where soils are richer, and *Sphagnum* bogs in wet pockets. Although the pure habitat is often fairly poor in species, it harbours some very useful plants – not least of which is heather itself. This wider patchwork with other habitat types creates a mosaic of upland vegetation we tend to lump together as 'heathland'.

Left: Bell heather by Sharon Tingey, watercolour (contemporary). This species (*Erica cinerea*) is often sold as 'lucky' Scottish heather. It is found on drier soils than our other species.

134 Chapter 5: Moorland and Mountains

Above the heath, on rocks, screes and barer ground, we find true alpine vegetation – crammed with specialised species that form mats or low whorls of leaves. Many species are remnants of the tundra that covered Scotland when the ice receded after the last Ice Age and which have crept their way up the mountains as the climate has generally warmed. Where they will go when the climate warms too much, and the mountains 'run out', is a major concern. As such, alpine habitats are a key focus of RBGE's conservation research and practical action because of the many rarities found there. Perhaps because of this rarity, few plants were used from true alpine areas, because reliable sources could not be found, or in many cases the plants were simply not known. Some species, however, are more common, finding their way to rocky and cliff-side habitats at all altitudes, and more of these were probably exploited.

We do have some tragic cases of lost species. At the expansion of the railways in the 1840s onwards, many day and weekend trippers would go to the hills, collecting plants as a hobby, with a particular premium placed on the rarities. In some cases, entire populations of now very rare plants were uprooted to help complete a collection.

Left: Sticky catchfly (*Silene viscaria*) by Mary O'Neil, watercolour (contemporary).
Right: Rock cinquefoil (*Drymocallis rupestris*) by Alexa Scott Plummer, watercolour (contemporary).
These rare mountain plants are both part of RBGE's Scottish Plant Conservation Programme.

Chapter 5: Moorland and Mountains

Ling heather *Calluna vulgaris*
Ericaceae (heather family)

English:	Dog Heather, He-heather, Heather, Ling
Scots:	Haddy, Hadder, Langa
Gaelic:	Fraoch
Flowering:	July to September
Habitat:	This species defines much of Scotland's heathland. There are two other relatively common heathers (*Erica tetralix* and *E. cinerea*), and several other similar members of the family in Scotland, but *Calluna vulgaris* is *the* iconic Scottish heather that covers hillsides and puts on an incredible flowering display in the autumn months. Calluna heathland is carefully managed through burning and grazing (by deer and sheep), which prevents many areas from returning to the forest they would otherwise be.

This perennial shrub grows to about 60 cm high, with tiny, triangular, scale-like leaves. The flowers are small, at 0.5 cm across, and form in spikes from pink or lilac to deep purple at the tips of the branches. The dry fruits contain tiny, dust-like seeds.

Heather was an invaluable construction material, and has been used as a thatch even to the present day. Ropes twisted from younger stems are incredibly durable, and sugen (or heather) rope was used to tie rafters and crucks together when making the supports for houses and roofs, as well as for tying the thatch itself down. Also used as a mattress, the whole plants, densely packed together in a box frame with the tops upwards, were said by Lightfoot (1777) to be, "not quite so soft and luxurious as beds of down, [but] are altogether as refreshing to those who sleep on them, and perhaps much more healthy".

The leaves give a yellow dye, with younger parts yielding a more vivid colour. Using alum as a mordant gives a deeper, orange shade, and the young tops yield a green. Although not as rich in tannins as oak bark or the tubers of tormentil (*Potentilla erecta,* in the 'Grasslands' chapter), the abundance of heather made it valuable as a tanning agent for leather.

Heather beer is a traditional brew, with recently revived versions now widely available. In the 1770s, the beer had fallen out of fashion, but a 'very potable' spirit was produced in Islay and Jura using two-thirds heather to one-third malt.

James Britten's *European Ferns* (1872) states that, "It seems that in some parts of Scotland it is a generally held belief that burning the heather will bring rain".

Left and below:
Ling heather by Fran Thomas, watercolour (contemporary).

Chapter 5: Moorland and Mountains 137

CLXXI. EMPETRACEÆ
EMPETRUM NIGRUM
ROYAL BOTANIC GARDEN, EDINBURGH

Crowberry *Empetrum nigrum*
Ericaceae

English:	Heather berry, Hillberry
Scots:	Berry hedder, Berry-girse, Crake berries, Crawberry, Craw croobs, Crawcroups, Croupert, Knauperts
Gaelic:	Lus na feannaig
Habitat:	Crowberry is usually found amongst heather

This perennial subshrub has slender, trailing woody stems, often creeping through heather. The leaves are short, and rolled-up into fat, blunt needles, with a distinctive white line on the underside. This is an adaptation to the dry, cold air of the tundra; crowberry was one of the first plants to recolonise the land after the glaciers had receded. The flowers are small at 0.5 cm across, and lilac to deep purple, and clustered in spikes among the leaves at the tips of branches.

One of Crowberry's main uses was as a dye. The juice of the berries was used to dye black or a range of purples, usually with alum (potassium aluminium sulphate). Lightfoot (1777) explains that crowberry 'juice' was used to produce a purplish-black dye (usually with alum) and that the berries give a black dye.

Left: *Empetrum nigrum* from the John Hutton Balfour teaching diagram collection (19th century).

This plant looks like some of the true heather species, and before Linnaeus invented the modern binomial system of plant names, one of its names was *Erica baccifera* – 'the berry-bearing heather'. Writing at the end of the 17th century, Martin Martin states that "*Erica baccifera*, boiled in a little water, and applied warm to the crown of the head and temples, is used … as a remedy to procure sleep."

The berries were eaten fresh but are often astringent and bitter so were made into jams. An unreferenced mention in the archives at the Royal Botanic Gardens, Kew relates that, "The fruits are said to be eaten by the peasantry." Lightfoot confirms this in his *Flora Scotica*, stating that Highlanders ate them, but he alleged they were not very tasty and bring on a headache if over eaten. Writing almost a century later, F. Buchanan White (1876) agreed, adding that he disliked the taste of crowberry jam.

Stag's-horn clubmoss
Lycopodium clavatum
Lycopodiaceae (clubmoss family)

Scots:	Fox fits, Tod's tails
Gaelic:	Lus a madhaidh-ruadh
Habitat:	Open moorland, sprawling through heather, but most noticeable on bare ground beside tracks and roadways

This herbaceous plant has stems sprawling up to 1 m through heather and on bare ground, rooting as they grow. The soft, needle-like leaves completely surround the stems. This plant has no flowers, instead producing pale cones around 5 cm high in summer. These grow to around 20 cm in total, with spore-bearing cones of 5 cm on the tips.

Stag's-horn clubmoss is a fascinating and versatile plant. The spores are fine and hydrophobic, which means they do not clump together, so it was used widely as a dusting powder for pills, latex gloves and condoms in Victorian and Edwardian times. The spores were also commonly used by stage magicians as a flash powder.

On a more local level, and in earlier times, Lightfoot (1777) mentions clubmoss (potentially a related species *Huperzia selago*) being used by Highlanders as an emetic and cathartic, but states that, "it operates violently, and unless taken in a small dose, brings on giddiness and convulsions". His contemporary, James Robertson, explains it is such a strong purgative that it can induce abortion, and that it was sometimes intentionally put to this use:

> the Girls when they happen to prove with child unmarried are aid to use a decoction of the Lycopodium selago in order to effect an abortion

Mary Beith's *Healing Threads* (2004) gives a very different use – steeping it in boiling water to give a soothing and softening lotion for girls' skin. It produced a red dye, and the spores were also used as a mordant instead of alum.

Alexander Carmichael's *Carmina Gadelica* (1900) extolls the supernatural virtues of clubmosses. He explains that if a traveller chances on clubmoss when out walking, they will not get lost. One of several poems and charms he records about clubmoss suggests a similar protection for travellers:

> The club-moss is on my person, No harm nor mishap can me befall; No sprite shall slay me, no arrow shall wound me, no fay nor dun water-nymph shall tear me.

Left and above: Preserved specimen of stag's-horn clubmoss, RBGE Herbarium, http://data.rbge.org.uk/herb/E00903009.

Chapter 5: Moorland and Mountains 141

Juniper *Juniperus communis*
Cupressaceae (cypress family)

Scots:	Aiten, Eaten, Eatin-berries, Jenepere, Melmet, Savin
Gaelic:	Aiteann
Habitat:	Found on heathlands or as an understorey shrub in Scots pine forest, Juniper is threatened by land clearance, cattle and a devastating disease caused by the fungus-like pathogen *Phytophthora ramorum*. Research by the Royal Botanic Garden Edinburgh is helping understand the disease and its spread, to ensure the future survival of juniper.

This shrub grows to 4 m – upright or spreading. The bark is reddish brown. The leaves are very sharp, dark-green needles with a distinctive fragrance when crushed. Male and female plants are separate, but both produce small cones; the female plant's cones develop into the juniper berries fundamental to gin or any good venison casserole.

Juniper is most famous as the key flavouring in gin, and stands of native juniper in Scotland were certainly harvested for their berries and wood. The wood burns with little smoke, so was ideal for fuelling illicit stills, without attracting the notice of the excise officers. As well as a flavouring, the berry-like cones were used as a stimulant and crushed to use as a poultice for treating snakebites. The renowned botanist W.J. Hooker, in his *Flora Scotica* (1821), explains they were widely used in the medicine of his day and that the wood was used for veneering.

Juniper's other well-known use was as an abortifacient, and there are tales of Mary Queen of Scots' maid, Mary Hamilton, trying to use it as such. The name 'mother's ruin', the nickname for gin, is widely known.

The ring ouzel bird (*Turdus torquatus*) is known in Scots as the Etenchaker, from the Gaelic aitean, and its migration route follows where juniper berries grow.

Right: Field illustration of juniper by Jane Wisely, watercolour (1943).

Cloudberry *Rubus chamaemorus*
Rosaceae (rose family)

Scots:	Averin, Everocks, Evron, Fintock, Knot, Knoutberry, Noup, Nub
Gaelic:	Lus nan oighreag
Flowering:	June to August
Habitat:	Upland heathland

This perennial herb has running rhizomes and aerial stems to 20 cm tall, with broad, round leaves with scalloped margins. The flowers are solitary and large at about 3 cm across, with white petals. The fruit is somewhat like an orange blackberry (bramble), but with fewer, larger drupelets.

The plant hails from upland blanket bogs across Scotland, and although widespread, it is seldom very common, producing few fruits. In the past it must have been plentiful enough because the seeds were found near the Iron Age Oakbank Crannog on Loch Tay and Lightfoot (1777) mentions their use as a dessert.

Cloudberries were used to make jams in more recent times, although the only place this can readily be found is in Ikea, which sells cloudberry jam imported from Sweden.

Johnson, writing in 1800, explains that the "the Scottish highlanders esteem the cloudberry one of their most useful and grateful fruits, especially on account of its long duration." However, Buchanan White writes 76 years later that "The taste of the fruit is very peculiar and (to my idea, at least) not very pleasant when uncooked; but when preserved, either as a jam or jelly, it is very agreeable and much sought after."

Right: Cloudberry from *Sowerby's English Botany*, Vol. 10 (1800).

144 Chapter 5: Moorland and Mountains

This deciduous shrub grows to 1.5 m tall. The leaves are dark green with silvery undersides and are covered in tiny golden spheres visible with the use of a hand lens. The leaf blades are wedge-shaped, with teeth towards the tip. There are separate male and female plants, which produce tiny flowers in cone-like catkins before the leaves flush out.

Bog myrtle was, and still is, used as a flavouring in soups and stews or as a garnish. The aromatic leaves work particularly well with lamb and venison. Johnston (1853) mentions their use in brewing beer, and this use has been revived in several modern craft beers.

Bog myrtle is often cited as an effective midge repellent and the terpenoid compounds it contains have been distilled out of the leaves by several companies in recent years. This insect repellent property perhaps gave credence to the idea it could keep away the faerie folk, and more prosaically it was used on Islay to keep flies out of the kitchen, and clothes moths away from fabrics. The smell is distinctive, and anglers often wear a sprig to keep the midges away, but the jury remains out on how effective it really is.

James Robertson (1770s) explains that, "The children [of Mull] are much troubled with worms for, which they use an infusion of the Myrica Gale or Foul."

Hooker (1821) suggests that when boiled in water the plant produces a waxy scum that could be used in candle-making. Its American relative *Myrica cerifera* was used for this on a commercial scale.

The leaves dye a variety of colours from yellows through to greens, and the stems are collected for flower arranging on a commercial scale.

Bog myrtle *Myrica gale*
Myricaceae (bog myrtle family)

English:	Scotch gale
Scots:	Foul, Gale, Gall-busses, Myrthus, Myrtle
Gaelic:	Roid
Flowering:	April and May
Habitat:	Wet heath, blanket bogs and grasslands on very acid, peaty soils

Left and below: Bog myrtle by Sarah Roberts, watercolour (contemporary).

Chapter 5: Moorland and Mountains   145

Below and right:
Scots pine by Elisabeth Scott, watercolour (contemporary).

Scots pine *Pinus sylvestris*
Pinaceae (the pine family)

English:	Deal (the wood)
Scots:	Banet fir, Bonnet fir, Bunnet fir, Burr, Moss fir, Pinule, Preenack, Sheepie
Gaelic:	Giuthas
Habitat:	Widely cultivated, but native woodland remnants are found in the Cairngorm National Park, and the north-west of Scotland

146 Chapter 5: Moorland and Mountains

This tree grows to 35 m. The platy bark is usually grey at the bottom of the trunk, with an orange flush towards the top. The needles are around 7 cm long, and borne in pairs. The male and female cones are separate but grow on the same tree. The males produce copious pollen in May and June, and the tiny, bright female cones take two years to mature when fertilised, eventually becoming the familiar pine cone.

This is Scotland's sole native pine tree. The species is thought to have been introduced to England by James VI (James I of England). James Robertson, writing in the 1770s, noted that there were few large trees left from once-extensive forests of very large trees; it was in constant demand as a timber tree for building in its native Scotland, suffering most heavily during the later Napoleonic wars. At the low point during the 1970s, only 10,000 ha were thought to remain. However, the seed bank remains in many places, and it is only grazing by sheep and deer that prevents the trees re-establishing. If heathland is fenced off for as little as 5 years, many young sapling Scots pines readily appear, and in 15 years there will be a forest of small trees with a rich herbaceous understorey.

Scots pine timber was known as deal, and used in many buildings for frames as well as for carpentry. It does not preserve well, but takes treatments such as pitch and varnishes, which will extend the lifespan. Although timber was the main product of Scots pine, it is a very versatile plant. The needles are used in beer-brewing even today. Historically, resinous roots were used for candles and the inner bark was used for making ropes, particularly in Ross-shire. It was also a major source of tannin, considered second only to oak.

Medicinally, the resin was mixed with pig fat and heated to make an ointment for treating boils, then placed on the sores for 12 hours. The astringent, tannin-rich bark was also used to treat fevers.

Chapter 5: Moorland and Mountains

Aspen *Populus tremula*
Salicaceae

Scots:	Esp, Old wive's tongues, Quakin ash, Tremlin tree
Gaelic:	Critheann
Flowering:	February and March
Habitat:	Damp woodlands, often near birch, and on heavy upland soils. Tolerant of acid conditions.

Above and right: Aspen by Linda Russell, watercolour (contemporary).

148 Chapter 5: Moorland and Mountains

This deciduous tree grows to 20 m. It produces many suckers, such that many stands of the trees are actually a single clone. The bark is silvery and smooth, with horizontal raised lenticels. The leaves are simple and round, with rounded teeth on the margins. The long, flexible pedicels cause them to tremble in the breeze. The flowers are pendulous catkins, present on separate male and female plants.

Hugh Fife, writing in the 1980s, explains that aspen was believed to induce prophetic visions, attested to in Robert Kirk's late-17th-century *Secret Commonwealth*, which discusses at length both prophecy and the faerie folk. Perhaps because of this, aspen was considered unlucky.

Lightfoot (1777) describes an 18th-century belief that the crucifix was made of the wood, evidenced by the fact that the leaves are always restless. Alexander Carmichael, writing 120 years later in 1900, confirms the continuation of this theory. Writing in his typically poetic style, Carmichael explains some of the aspen's ill reputation:

> The people of Uist say that the hateful aspen is banned three times. The aspen is banned the first time because it haughtily held up its head whilst all the other trees of the forest bowed their heads lowly down as the King of all created things was being led to Calvary. And the aspen is banned the second time because it was chosen by the enemies of Christ for the cross upon which to crucify the saviour of mankind. And the aspen is banned the third time because … [here the reciter's memory failed him]. Hence the ever-tremulous, ever-quaking motion of the guilty, hateful aspen even in the stillest air. Clods and stones and other missiles are hurled at the aspen by the people. The reciter, a man of much natural intelligence, said that he always took off his bonnet and cursed the hateful aspen in all sincerity wherever he saw it. No crofter in Uist would use aspen about his plough or about his harrows, or about his farming implements of any kind. Nor would a fisherman use aspen about his boat or about his creels or about any fishing gear whatsoever.

Above:
Rowan by Nicola Macartney,
watercolour (contemporary).

150 Chapter 5: Moorland and Mountains

This small deciduous tree can be single, or few stemmed, with bronze, smooth bark when young, ageing to a dull silver. The lenticels are large and horizontal. The leaves are compound yet feathery, with a terminal leaflet; the leaflets are downy, paler underneath, and with large teeth on the margins. The flower heads are a dense corymb of creamy white flowers. The berries are a vivid red.

Rowan is one of the classic occult plants in Scotland; it appears in many charms and stories connected with witchcraft and the Sidhe (faerie folk). Rowan wood was particularly useful in warding off the effect of the evil eye, or preventing milk from being spoiled or stolen by malign magic. The flour for ceremonial cakes was made using a threshing tool made of rowan, called the buaitean in Gaelic. Funeral biers were often made of rowan wood.

'Magic hoops' of plant stems were often woven and placed under milk storage jugs, pails and churns to prevent the milk from being stolen. As well as rowan, ivy (*Hedera helix*) and a plant called 'woodbine' (probably honeysuckle, *Lonicera periclymenum*) were used as a trinity of plants to make the hoops. This practice ran through to the 19th century at least, but the treatise on *Daemonologie* by James VI (James I of England) (1597) expresses scepticism, debating whether or not "commonlie daft wives … knitting Rowan trees" to protect livestock were dabbling in true deviltry.

An interesting little anonymous verse goes, "Black luggie, Lammer [amber] bead, Rowan-tree, and red thread, Put the witches to their speed." Black luggie may be the 'Jew's ear' *Auricularia auricula-judae*, a common fungus found on elder trees, or a pail made of staves with one projecting above the others as a handle to allow it to be lugged around. An alternative version is, "Rowan tree and red thread gar the witches tyne their speed."

In his travels around Scotland, Lightfoot (1777) noticed that a disproportionate number of rowans were to be found near

Rowan *Sorbus aucuparia*
Rosaceae (rose family)

English:	Mountain ash
Scots:	Quicken, Raun, Rodden, Rone berry, Rone tree, Roddin-tree, Rountree
Gaelic:	Caorann
Flowering:	May and June, fruiting around two months later
Habitat:	Widespread in mixed and deciduous woodland, and as solitary specimens in exposed sites on moorlands and hills. The plant can cling on in precarious crevices and gullies.

standing stones. He suggested that these may have been remnants of plantings by druids. He also mentions that shepherd girls would usually drive the flock with a rowan stick, and that in Strathspey livestock were made to pass through a hoop made of rowan in the morning and evening, as a charm against black magic.

Rowan's uses were not all supernatural. In passing, Lightfoot also notes the use of the berries for distilling. Mary Beith (2004) explains how the berries were boiled up in water until they pulp down, strained through a muslin and used as an effective gargling solution. Rowanberries and apple, with a little brown sugar, were made into a decoction and used to relieve whooping cough. The wood is good, and J. D. Hooker (1821) notes that it was valued for its compactness and that among other things it was used to make whistles.

Even now, many modern houses have a rowan tree in the front garden, supposedly a sovereign barrier to witches, and rowan has been taken to New Zealand, where the tradition carries on.

Chapter 5: Moorland and Mountains

Blaeberry
Vaccinium myrtillus
Ericaceae (heather family)

English:	Bilberry, Huckleberry, Whortleberry
Scots:	Blairdie, Blivert
Gaelic:	Caora-mhiteag
Flowering:	April to July, fruiting around 2 months later
Habitat:	Upland heath and Scots pinewood

Left: Field illustration of blaeberry by Jane Wisely, watercolour (1939).

Chapter 5: Moorland and Mountains

This deciduous perennial subshrub grows to 50 cm. The leaves are small and ovate with tiny teeth on the margins. The flowers are small, urn-shaped, and pale cream or with a pink tinge. The fruit are like a small blueberry.

Blaeberry was collected informally when out in the hills, or in much greater quantities for market, using metal-tined combs to pull the berries from the plants. Buchanan White (1876) lists it among very few fruits collected in sufficient quantities to bring to market, stating that a "considerable amount of money is made by those who gather them for sale". Blaeberry Hill, near Eskdalemuir in Dumfriesshire, is named for the berries collected for market from there. The berries were eaten fresh or made into a jelly for eating with game, and this was often dissolved in whisky, as Lightfoot puts it, "to give it a relish to strangers". Lightfoot (1777) also describes it as a "cooling and agreeable food".

Blaeberry is a classic example of a popular food that was seen to cross over into medicine. Lightfoot mentions its use in treating diarrhoea and dysentery, for which Martin Martin (1703) says blueberry syrup was used 75 years earlier. Mary Beith (2004) records its use as an infusion in treating pain.

The berries produce a violet dye, and blue with copperas (iron sulphate). With nut galls, it gives a brown dye, and Verdigris (copper) and sal ammonia, a red. Modern dyers report that blaeberries yield a pink-purple (with alum), green (with alum with ammonia rinse), grey (with alum and iron), and a slate blue (frozen berries with alum and cream of tartar).

The related lingonberry, also known as cowberry (*Vaccinium vitis-idaea*), is not recorded in so many uses in Scotland but remains very popular in Scandinavian countries, where it is widely used as a flavouring imported from Sweden via Ikea. The fruit is more tart, and it is somewhat less common than blaeberry, but Buchanan White (1876) mentions it and cranberries being made into a "capital jelly".

Above: Field illustration of lingonberry by Jane Wisely, watercolour (1968).

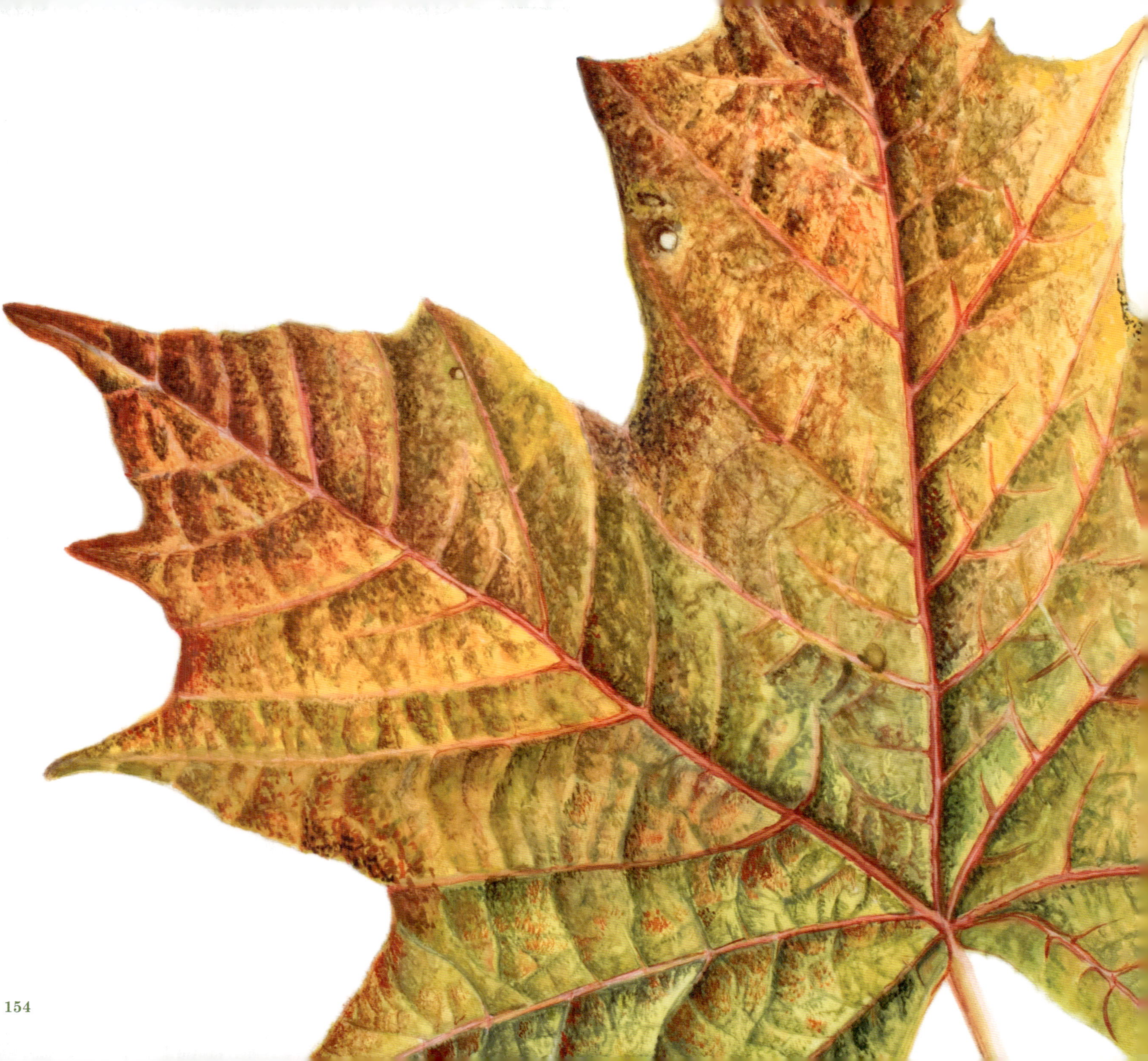

Chapter 6
Human Habitats

Urban areas and other heavily disturbed habitats have their fair share of useful plants, from the humble native nettle (*Urtica dioica*) to suspected Roman or medieval introductions such as ground elder (*Aegopodium podagraria*) and sweet cicely (*Myrrhis odorata*). These last two are now fairly common woodland residents, perhaps too common in the case of ground elder. Most of the plants in this chapter belong in wild habitats too but are specialised ruderals, often weedy plants with fast growth and reproductive cycles. Ruderals like these are usually from disturbed habitats such as streamsides, open rocky and sandy soils that are overgrazed, similar in many ways to the brownfield sites they are found in nowadays. Others come from areas of very rich nutrients similar to our over-fertilised urban and arable landscapes. These plants have all taken the opportunity the effects of human activities on the landscape and are perhaps far more successful than they would 'naturally' be. For good or ill, we make a huge impact on our environment, but it is fascinating to see the myriad ways in which plants and their communities respond.

Left: Sycamore by Janis Bain, watercolour (contemporary).

Burdock *Arctium minus*
Compositae/Asteraceae (daisy family)

English:	Flapper-bags (a name shared with butterbur, *Petasites hybridus*)
Scots:	Bardog, Burdocken, Burrs
Gaelic:	Leadan liosda
Flowering:	July to September
Habitat:	Hedgerows and woodland margins, especially by paths and roadsides, where the fruits are dispersed by unwitting animals. They are also common on disturbed ground such as shale bings.

This robust, very deep-rooted biennial herb grows to 1.3 m tall. The lower lance-shaped leaves grow to 30 cm long, and are very downy. The many tiny flowers are clustered together in a 2-cm spherical, thistle-like head. The flowers have pink petals and white styles projecting from the head, which is covered in bracts tipped with recurved, hook-like spines (the inspiration for Velcro). When dry, these fruit heads attach to passing animals and the fruits within are shaken to the ground.

Burdock root is used for dandelion and burdock drink. It is dug up, sautéed and eaten by wild foragers as a form of *gobo*, the popular Japanese pub food.

Its most fascinating use, however, is as part of the Burry Man's costume. South Queensferry's Burry Man parade is an annual event whose origins are hazy; it was first recorded in the mid-1600s. On the second Friday of August each year, a local man dresses in a costume covered head to toe in the burrs of *Arctium minus*, which are collected from byways and shale bings locally. The Burry Man, as the character is known, walks around the streets of the town, from around nine o'clock in the morning. As he makes his way around South Queensferry, he calls at each of the pubs in the town (as well as some other locations) where he receives a glass of whisky. He is accompanied by two attendants, who ensure the ordeal is endurable. Even so, by the time the Burry Man has completed his circuit (after some nine hours of slow walking), he will have had a considerable amount to drink and be exhausted by the weight of the costume. The Burry Man is seen as either a classic representation of the 'Green Man' of British folklore (a nature and vegetation spirit) or else as a 'scapegoat' figure, carrying the malignant spirits or sins of the townspeople away on his burrs. Fortunately, considering the scapegoat scenario, nowadays the Burry Man is not sacrificed or driven from the town as he might have been in the ceremony's earliest days. One other theory is that the event commemorates the landing at the town of Queen Margaret, wife of Malcolm Canmore. There is some difference of opinion among the local children as to how scary the Burry Man is. To some he is a comic figure; to others, the first sight of the Burry Man in childhood is enough to "scar a person for life". Some of the town's children dole out the following sage advice: "Dinnae look in his eyes, he'll send ye mad."

Left and above: Preserved specimen of lesser burdock, RBGE Herbarium, http://data.rbge.org.uk/herb/E00903164.

Chapter 6: Human Habitats

158 Chapter 6: Human Habitats

Daisy *Bellis perennis*
Compositae, Asteraceae (daisy family)

Scots:	Bairnwort, Benner gowan, Curly-doddy, Daseyne, Ewe-gowan, Golland, Gowan, Kokkelootie, Mary gowlan, May gowan, Wallie
Gaelic:	Neoinean
Flowering:	March to October
Habitat:	Grasslands, parklands and garden lawns

This perennial, rosette-forming herb grows to 10 cm tall. The leaves are short and spoon-shaped, with rounded teeth on the margins, and reach 5 cm long. The flowers are tiny; the central ones are yellow and outer ones are white and tipped with pink, with larger ray petals.

Maud Grieve (1931) explains one of this plant's names through its most popular use: "In Scotland it is the 'Bairnwort,' testifying to the joy of children in gathering it for daisy-chains." Many archival records mention the hugely versatile little flowers used as ingredients in children's imaginary cooking games. Aside from this, these instantly recognisable little plants have been used medicinally.

As well as treating toothache (for more on this, see under Yellow flag in the 'Wetlands' chapter), they were also used for treating eye problems. This use may be the result of following the Doctrine of Signatures, in which other plants with vivid, open little flowers such as the appropriately named eyebrights (*Euphrasia* species) were used because of their perceived similarity to an eye. Mary Beith (2004) also mentions them as a constituent of ointments to treat cuts and bruises, and it seems that this practice was particularly prevalent on Colonsay.

Roots and leaves were used for rheumatism and for gout. Our anonymous record for this says that one pint of boiling water was poured over the leaves and then it was cooled and strained, and was taken three times daily (presumably as a drink).

Alexander Carmichael (1900) records daisy and burdock (discussed elsewhere in this chapter) in a supplication to St Magnus of Orkney: "Sprinkle dew from the grass upon the kine, Give growth to grass, and corn, and sap to plants, Watercress, deer's-grass, ceis, burdock, and daisy." Ceis is a Gaelic word for a spear, or lance and may refer to any of a multitude of 'spear-like' plants.

Left: Field illustration of daisy by Jane Wisely, watercolour (1975).

Hemlock
Conium maculatum
Apiaceae/Umbelliferae (carrot family)

Scots:	Bunnel, Coo-cakes, Hech-how, Humloik (variant spellings), Humly rose, Kaka, Kex, Scab
Gaelic:	Iteodha Many of the names given to hemlock also apply to other umbellifers; they can be very difficult to distinguish from each other, as testified by a tragic Victorian poisoning case in which a father foraging on Arthur's Seat for 'wild carrots' made a fatal error of identification, and consumed hemlock instead.
Flowering:	June and July
Habitat:	Often dry ground, especially by paths and roadsides. Much like with cow parsley (*Anthriscus sylvestris*), roads are a major route along which hemlock travels and botanists are tracking its perceived advance northwards.

Left: Field illustration of hemlock by Jane Wisely, watercolour (1969). Note the leaves are illustrated opposite one another, where they are usually alternate or spiral up the stem in actuality.
Right: Field illustration of hemlock by Jane Wisely, ink (1969).

This biennial herb grows to 1.5 m tall. It is hairless and slightly greyish green throughout, but with distinctive purple blotches on the lower parts of the hollow stems. The leaves are dissected into fine leaflets. The flower heads are umbels with small white flowers. The whole plant smells unpleasantly musty; the scent is often likened to that of mouse urine.

Humloik, a Scots name for hemlock, is an excellent example of an etymological problem, discussed in John Wood's *Scotch Names of Wild Flowers* (1893). It has been suggested that the names is derived from the Saxon (as with many Scots words) hem (edge or border) and leac (leek), meaning a leek-like plant from the edge of cultivated land. This seems appropriate enough, but it could equally well be applied to a huge number of plants. It is hemlock's notorious toxicity and sinister reputation that one might expect to distinguish it from other plants, and Wood suggests that the Scots name is, in fact, derived from these properties. 'Haulm' approximates to a stalk, or stem (typically applied to cereals) and 'lyke', or 'lic', refers to a corpse, therefore hemlock would be 'corpse plant' or 'stem of death'. As with many attempts to trace the etymology of plant names, however, much is speculative.

Despite its deserved ill-reputation, hemlock was used medicinally. William Jackson Hooker's *Flora Scotica* (1821) explains that it is, "highly narcotic and dangerous in large quantities: has been much used medicinally in ulcerous and cancerous disorders". Towards the end of the same century, Alexander Carmichael (1900) writes about the same use:

> A cold poultice of hemlock was applied to a cancer sore. The hemlock plaster was so hot and so strong that it drew the cancer out from the bottom, the roots coming with the cancer as the roots come with the hemlock itself out of the ground. This was effective when done in time. When the disease became soft nothing could cure it. A police inspector in Glasgow had cancer in the lower lip. Instead of having it cut he went to his native home near Fort William. There a man applied a poultice of hemlock to the lip and extracted the cancer bodily. The patient told my informant that the pain had been excruciating. The flesh containing the cancer came away, carrying the roots with it. These were very numerous and resembled, the fine thready roots of hemlock, he could hear the sound of their breaking away. The patient placed the cancer in spirits and kept it for many years. My informant said that except for a hollow the man's lip seemed quite healthy and normal.

Chapter 6: Human Habitats 161

Common ragwort
Senecio jacobaea and Groundsel *S. vulgaris*

Asteraceae/Compositae (daisy family)

English:	*S. jacobaea*: Ragweed, Yellow weed
Scots:	*S. jacobaea*: Benweed/Bunweed, Fizz-gig, Gollan, Stinking Alisander/Elshender, Stinking Davie, Stinking Willie, Wee-bo/Weebies
	S. vulgaris: Grundsel, Gruniswallow, Swally, Wattery drums
Gaelic:	Buaghallan
Flowering:	June to October for ragwort, and year-round for groundsel
Habitat:	Open ground: ragwort tends to favour meadows, and grazed areas, where it can take over huge areas. Livestock prefer to avoid it, and it is potentially very toxic to most animals. Groundsel is a classic weedy plant – popping up anywhere it can – from cracks in urban pavements to forestry tracks.

Ragwort is a robust biennial herb to about 1 m tall. The leaves are finely dissected and hairless, with blunt and rounded tips to the leaflets. The flower head is a group of many composite, daisy-like capitula, each up to 2.5 cm across, with tiny yellow-petalled central florets and yellow ray florets. The plant smells strong, and fairly unpleasant. Groundsel is annual and looks

Chapter 6: Human Habitats

like ragwort's weedier cousin: shorter, with smaller leaves and nodding flower heads lacking obvious ray florets. Groundsel often has a strong red tinge to the stem; this is less obvious with ragwort.

Because of its invasive habit and toxicity to livestock, ragwort has a terrible reputation as a weed, which is nicely captured in the account by Alexander Carmichael (see under Docks in the 'Wetlands' chapter).

Faerie folk were said to shelter under the leaves, and the stems were apparently rode by faeries and witches abroad at night. Interestingly, Martin Martin (1703), mentions groundsel roots being used as an amulet to prevent the cream being stolen from the milk.

Ragwort was by no means all bad. Where material for baskets was scarce, on the Western and Northern Isles, ragwort stems were a valuable substitute for willow. It was also a valued source of many shades of yellow and orange dyes, domestically and for the Harris Tweed industry, although the dyeing process apparently smells foul. The strong smell of the plant was useful in keeping mice away from clothes or food, somewhat like tansy (*Tanacetum vulgare*), a close relative of ragwort.

Medicinally, groundsel was used in Martin Martin's time (late 1700s):

> To ripen a tumour or boil they cut female jacobea small, mix it with some fresh butter on a hot stone, and apply it warm; and this ripens and draws the tumour quickly, and without pain; the same remedy is used for women's breasts that are hard or swelled.

Martin Martin (1703) also mentions groundsel being used as a strong emetic and to help cool people. Lightfoot and Robertson, both writing in the 1700s, record groundsel being used as a cataplasm to draw pus from boils. Each of these writers was trained in botany, so their identifications are likely to be correct.

Left: Ragwort from *Sowerby's English Botany*, Vol. 16 (1803).
Right: Groundsel from *Sowerby's English Botany*, Vol. 11 (1800).

Chapter 6: Human Habitats 163

Red clover *Trifolium pratense* and white clover *T. repens*

Leguminosae/Fabaceae (pea family)

Scots:	Clever, Cow-cloos, Cow-grass, Curl-doddy, Plyvens, Soukies, Soukie-soo, Sucklers, Triffle
Gaelic:	*T. pratense*: Seamrag Dhearg *T. repens*: Seamrag Bhan
Flowering:	May to September
Habitat:	Fairly ubiquitous: red clover is typical in richer meadows, and white clover in a wider range of habitats, including parks, lawns and disturbed ground

This creeping herbs grows to 30 cm tall (although *T. repens* is smaller). The leaves are a characteristic trefoil, often with white chevron markings on each of the three leaflets. The flowers are tiny pea flowers in dense heads. These two species' common names of red clover and white clover are appropriate for their flower colours.

There are several other similar clover species, but these are the two most common. It is likely that very closely related species were not distinguished, and were used in similar ways.

There is some debate as to whether the true shamrock is one of the clovers (*Trifolium*) or wood sorrel (*Oxalis acetosella*). Both have three leaflets per leaf, which was used to illustrate the Holy Trinity, but this appearance has since become somewhat muddled with the idea of the lucky four-leaved clover. *Trifolium* plants often have leaves with four or more leaflets, but there is no evidence of a 'four-leaved wood sorrel'. Alexander Carmichael's verses collected at the end of the 1800s support this confusion, mentioning "the four leaves of the straight stem", when talking about clover in a short spoken charm that includes many Christian allusions.

Both Geoffrey Grigson (1955) and Roy Vickery (1997) record an interesting belief. This is from the School of Scottish Studies manuscripts and explains that when a foal is born, it expels a "dubhliath [which looks like cormorant or rabbit's liver, about the size of a crown coin] from its nostril. [The informant] kept one to prove its existence to young people. If it is kept for seven years, a four-leafed clover will grow from it". Some accounts suggest it should be buried and a 'long time after' a four-leaved clover will grow on the spot.

This plant has long been used in crop rotations, because the root nodules harbour *Rhizobium* bacteria that help to fix nitrogen in the soil. The closely related white clover (*T. repens*) was an impromptu snack for children – the heads would be plucked and the nectar sucked from them.

Right: Red clover by Marie Barbour, watercolour (contemporary).

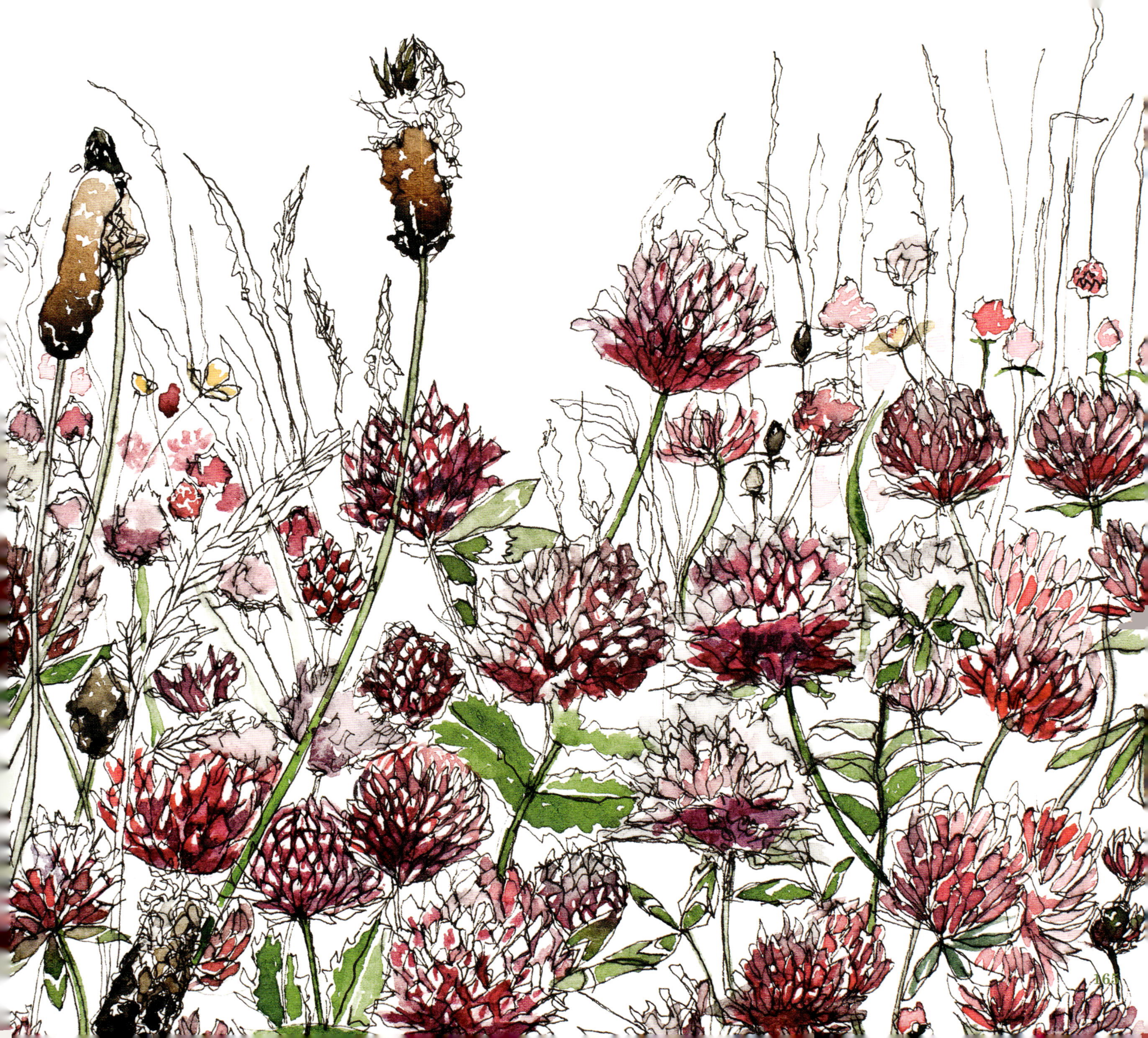

Nettle *Urtica dioica*
Urticaceae (nettle family)

Scots:	Heg-beg, Jag, Jinnie nettle, Jobbie nettle
Gaelic:	Deanntag
Flowering:	June to August
Habitat:	Open woodlands and disturbed sites – a common urban plant

Main and right: Nettle by Gloria Newlan, watercolour (contemporary).

Nettles are perennial herbs to 1.5 m tall (usually less than 1 m), spreading by underground rhizomes. The leaves grow in opposite pairs up the square stem (decussate), to 10 cm long, with teeth on the margins. There are stinging hairs throughout. The female and male plants are separate, with the flowers on both sexes very small, and borne on green catkins.

Nettles are one of several plants used as a tonic in various forms. Widely used even today by foragers as a pot herb, the young plants or tops are gathered and used much like spinach. As a soup, or mixed in with porridge, many areas had their own version of what was sometimes termed 'St Columba's broth', or nettle broth (cal deanntaig). The tradition in early spring was to take a set number of meals of the broth.

Martin Martin (1703) records nettle leaves, crushed and added to meat or lentil broth, being used to alleviate rheumatism, and on Lewis these were fermented with reeds to produce an expectorant. Interestingly, an account almost 200 years later in the 1880s mentions a similar nettle ale being used to treat jaundice. Treatments for rheumatism and muscular pain appear throughout many sources over 350 years.

Lightfoot (1777) mentions nettles being made into made into a strong decoction with salt, as a rennet substitute to curdle milk on Arran, often with *Galium verum* (lady's bedstraw).

The traditional idea that nettles grow in the slightly more alkaline areas where people urinated outside the house carries through today to explain the patches of nettles often seen near ruined blackhouses. However, as Johnson writes in his *Useful Plants of Great Britain* (1862), nettles were often semi-cultivated, and before his time it was common practice in Scotland to "Force the early nettles for their spring kail". Johnson also mentions a wonderful report from a Mr Campbell who had eaten nettles:

> I have slept in nettle sheets and I have dined off a nettle table-cloth. The young and tender nettle is an excellent pot-herb. The stalks of the old nettle are as good as flax for making cloth. I have heard my mother say, that she thought nettle more durable than any other species of linen.

The versatile nettle was used for dyeing as well: green from the leaves, yellow from the roots and black with iron as a mordant.

Chapter 6: Human Habitats

Dandelion
Taraxacum officinale
Asteraceae/Compositae (daisy family)

Scots:	Bitter aks, Bumming pip/Bum-pipe, Devil's milk plant, Doon-head clock, Eksis girse, Horse gowan, Milk gowan, Pisstebed/Pish-the-bed, Stink Davie, What-o-clock-is-it, Witch gowan, Yellow gowan
Gaelic:	Bearnan Bride
Flowering:	Year-round
Habitat:	Ubiquitous – this plant is common in disturbed areas almost everywhere, with many microspecies found in local areas.

This perennial herb grows to 15 cm tall. It has a deep taproot, with soft, saw-toothed leaves forming a rosette. The flowers are very small and numerous, with bright gold petals borne on an often red-tinged peduncle.

F. Marian McNeill's fascinating *The Silver Bough* (1957) explains that there are a number of plants to which magical properties of milk protection were ascribed. 'Magic hoops' of plant stems were woven and placed under the milk storage jugs, pails and churns to prevent the milk from being spirited away by the Sidhe (faerie folk). One hoop was made from milkwort (*Polygala vulgaris*), butterwort (*Pinguicula*

Left and right: Dandelion by Gloria Newlan, watercolour (contemporary).

vulgaris), dandelion (*Taraxacum* spp.) and marigold (*Calendula officinalis*, an introduced species), bound with three threads of lint (from *Linum catharticum*, the fairy flax) and placed under the container while the Holy Trinity was invoked.

The leaves are eaten in salads, although there was always the perceived danger that touching or even smelling them would cause children to wet the bed. This belief is common in many north European countries, where the common names reflect the plant's diuretic properties. Indeed, it was used historically as a diuretic, and still is amongst herbal medicine practitioners.

Arguably an even more famous belief is that the plant can tell the time, as reflected in the name 'What-o-clock-is-it?'. The number of puffs of breath that it takes to disperse all the little parachuted fruits from the fruiting head supposedly tells you what hour of the day it is. Once you have finished telling the time, the remaining stalk also has its uses. The peduncle is a tube and can be pressed into good service as a pipe, making a humming noise like a kazoo.

This gives the plant yet another odd-sounding name: bum-pipe. In this case, 'bum' refers to hum, rather than anything ruder. These pipes were also used as a substitute for bike tyre valves during World War II on Royal Air Force bases.

The best coffee substitute was considered to be the roasted and powdered roots of dandelion, which were thought to help alleviate an upset stomach, and in the Glencoe area in recent times dandelion sandwiches were considered good for stomach ulcers. Rubbing the juice on warts was said to cure them.

Minor species in urban areas

Many native ruderal species have found homes in and around our streets and the edges of parks. Some, such as mugwort (*Artemisia vulgaris*), were certainly cultivated, and in this case used medicinally and as a flavouring in beer, as the name suggests. Lightfoot (1777) mentions its use as a pot herb, and as in many cases, the food was also thought to have a medicinal value. This is recorded in a song passed by Rodney Higgins to Roy Vickery and included in the latter's *Dictionary of Plant Lore* (1997): "if they wad drink nettles in March and eat Muggins in May, Sae Many braw young maidens would nae be going to clay."

Carrying mugwort was believed to ward off tiredness, and it was smoked by those who could not afford tobacco. Coltsfoot (*Tussilago farfara*) was perhaps the most famous tobacco substitute, and the leaves were smoked and also, strangely, billed as being *good* for asthma! It was also given to relieve a dry cough, almost Europe-wide in the form of juice, fresh leaves or a syrup.

Chickweed (*Stellaria media*) is a popular wild-foraged food today, but Martin Martin (1703) explains an alternative use:

> To procure sleep after a fever, the feet, knees and ankles of the patient are washed in warm water, into which a good quantity of the chick-weed is put, and afterwards some of the plant is applied warm to the neck, and between the shoulders, as the patient goes to bed.

Hogweed (*Heracleum sphondylium*) has an undeserved notoriety because of its larger cousin – the giant hogweed, an invasive non-native species that contains a group of chemicals called psoralens, which can cause acute photoreactive dermatitis.

Left: Hogweed by Sheila Anderson Hardy, ink wash (contemporary).
Right: Maidenhair spleenwort by Jan Miller, watercolour (contemporary).

The native hogweed is smaller and still contains relatively small amounts of the chemicals. It was put to use as a fodder plant, and in brewing, and the tender shoots were eaten raw. DNA research by Jin-Hyub Paik of the Royal Botanic Garden Edinburgh has revealed why this might be: hogweed and parsnip are more closely related than we first supposed.

Children in particular come up with some superb names, and inventive uses for plants. *Plantago lanceolata*, the ribwort plantain – is a great case in point. The flower heads were used in a game like conkers, with one battler holding their flower out while the other tried to whip the head off it with theirs. Common names for the plant reflect this sport of playground warriors and include battlers, fechters (fighters), kemp (to fight), sodgers (soldiers), and for those who listened in history class, Carl-Dodie (Bonnie Prince Charlie versus George II).

Buttercups have adapted well to urban grasslands. *Ranunculus acris* (meadow buttercup) is found in wetter areas, whereas *R. repens* tends to be more weedy. The irritant sap of these and other species were often used to raise blisters, which were drained to help balance the humours in traditional medicine.

Martin Martin (1703) mentions two common urban ferns in an intriguing recipe. The maidenhair spleenwort (*Asplenium trichomanes*) and hart's tongue (*A. scolopendiurm*) were boiled in wort (the infusion of malt or barley prior to fermentation), and the resulting beer was drunk to combat coughs or consumptions. Lightfoot (1777) reports that, "The country people sometimes give a tea or syrup of it for coughs and other complaints of the thorax, but it is rarely used in the shops". He also mentions maidenhair spleenwort being used as a vulnerary to help heal wounds and burns.

Chapter 6: Human Habitats

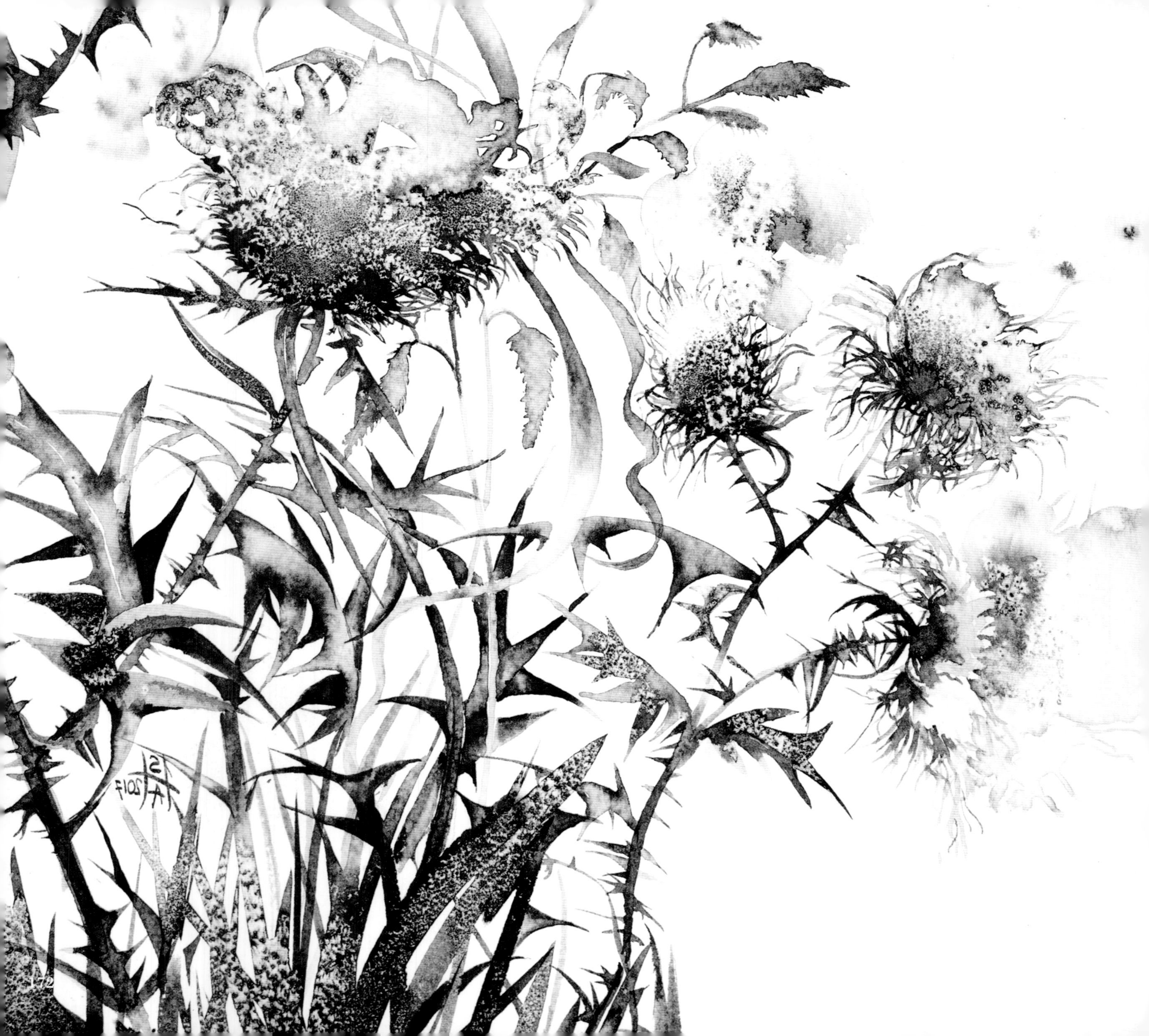

A final note on Scotland's national plant

The thistles belong to several closely related groups; *Cirsium*, *Carduus* and *Onopordon*. Quite which species amongst them is the true Scottish thistle has long been a source of debate, with most of the discussion focused on two species: spear thistle (*Cirsium vulgare*) and cotton thistle (*Onopordon aanthium*). The former is impressive, common and native, whereas the cotton thistle is even more regal but was introduced from the Mediterranean as an ornamental. *Onopordon* was the possible inspiration for the heraldic device of the Stewart kings, and was also the species Walter Scott asked to be carried for the visit of George IV to Edinburgh in 1822. Heraldic thistles in Scottish iconography tend to look a little more like the cotton thistle, but this is not always consistent. It does not, however, appear in *Hortus Medicus Edinburgensis* (1683) as growing in the Botanic Garden in the late 1600s. The only thistle mentioned is *Silybum marianium*, the Lady Mary's Thistle, or milk thistle, widely used medicinally at the time.

The debate was largely put to bed by Agnes Walker, working with Jim Dickson, who concluded that *Cirsium vulgare* is the likely native inspiration, but took the eminently sensible view that other species may be considered the Scottish thistle depending on the context. My personal favourite lies very slightly with the beautiful Mediterranean species. Also, the taxonomic name *Onopordon* means something along the line of 'donkey's fart' – an exquisitely romantic name for such an iconic plant.

Left: Spear thistle by Sheila Anderson Hardy, ink-wash (contemporary).

Right: Scotch or cotton thistle by Clare McGhee, watercolour (contemporary).

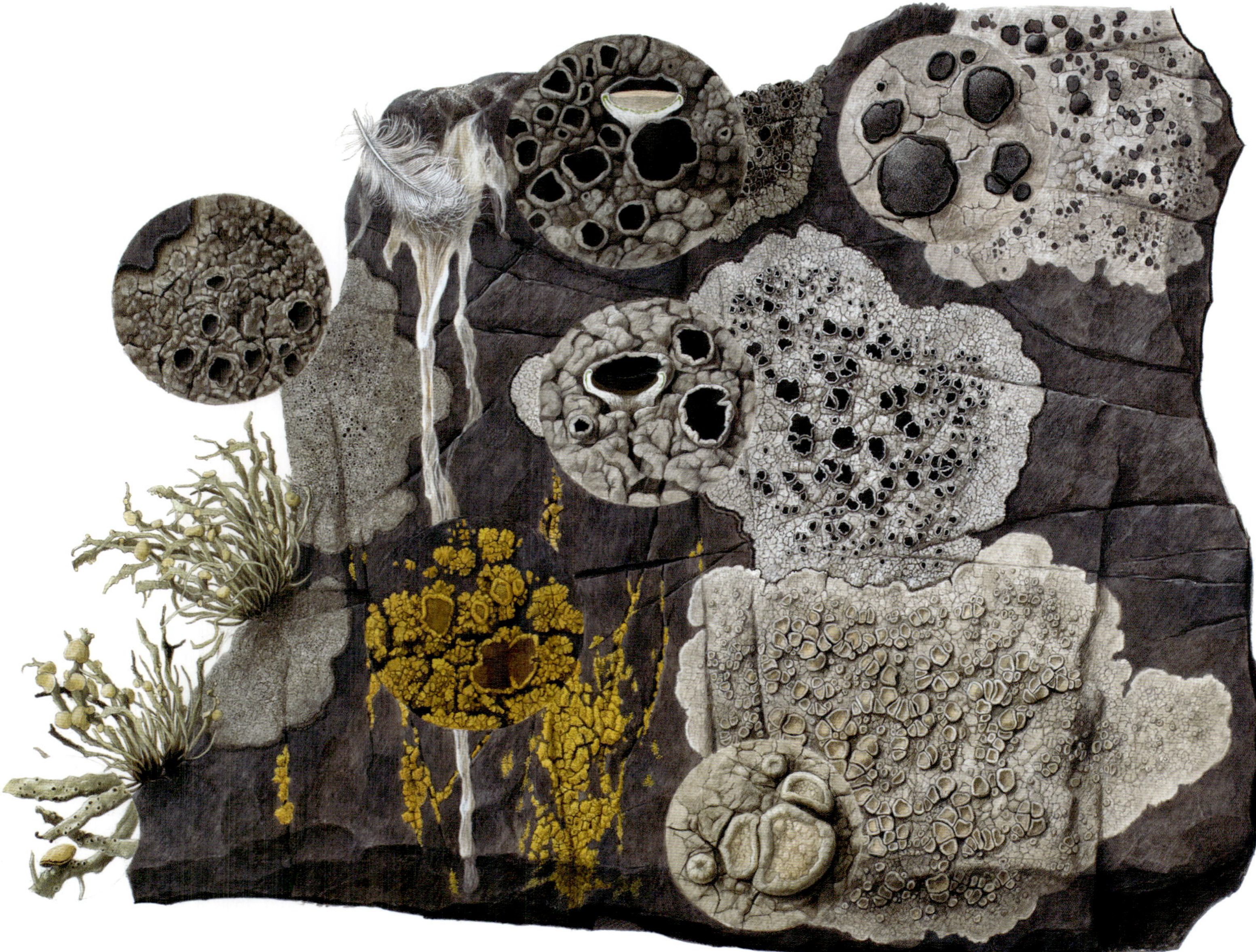

Glossary

Adventitious
Added or appearing accidentally

Alum
A double sulphate of aluminium and another element, in particular phosphate

Aril
An extra seed covering, typically coloured and hairy or fleshy

Basal
Forming the base

Cataplasm
A poultice or plaster

Copperas
Iron sulphate, mostly used for the fixing of wool colours

Corymb
A flower cluster whose lower stalks are proportionally longer so that the flowers form a flat or slightly convex head

Cyme
A flower cluster with a central stem bearing a single terminal flower that develops first, the other flowers in the cluster developing as terminal buds of lateral stems

Decoction
Essence extracted from a plant by boiling

Deobstruent
A medicine taken to remove obstructions

Drupe
A fleshy fruit with thin skin and a central stone containing the seed

Drupelet
Any of the small individual drupes forming a fleshy aggregate fruit such as a blackberry or raspberry

Emetic
A substance that induces vomiting

Lanceolate
Shaped like a lance head; of a narrow oval shape tapering to a point at each end

Lenticels
Raised pores in the stem of a woody plant that allows gas exchange between the atmosphere and the internal tissues

Mordant
A substance used to fix dyes

Palmate
Leaves that have five or more lobes whose midribs all radiate from one point

Pedicels
Small stalks that bear an individual flower in an inflorescence

Peduncle
A stalk bearing a flower or fruit or the main stalk of an inflorescence

Petiole
The stalk that joins a leaf to a stem

Pinnate
Leaves that have leaflets arranged on either side of the stem, typically in pairs opposite each other

Raceme
A flower cluster with the separate flowers attached by short equal stalks at equal distances along a central stem: the flowers at the base of the central stem develop first

Rachis
A stem of a plant, especially a grass, bearing flower stalks at short intervals

Sporangia
Receptacles in which asexual spores are formed

Sudorific
A medicine that induces sweating

Left: *Lecanora gangaleoides, Tephromela atra, Porpidia albocaerulescens, Aspicilia leprosescens, Candelariella vitellina, Ochrolechia parella*: original illustrations x 2.4 and x 12, Ramalina cuspidata x 1.8 and x 6, by Claire Dalby RWS RE, watercolour (contemporary).

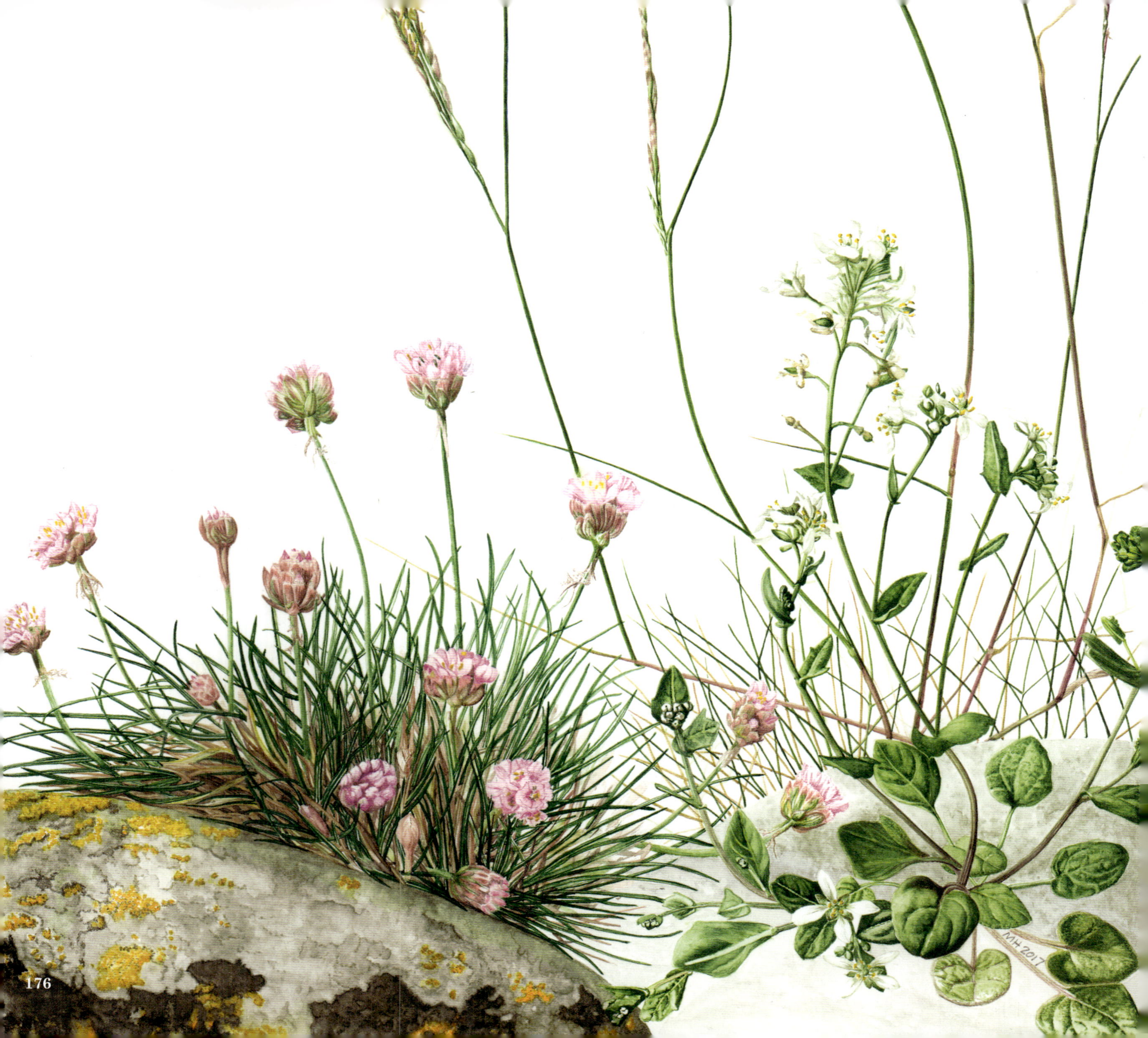
MH 2017

Bibliography

In addition to the literature discussed in the introduction, these are the other sources mentioned throughout the text. Of course, there are many other fascinating books, articles and snippets available to the interested reader.

Buchanan White, F. (1876). The edible wild fruits of Scotland. [Part 1] *Scottish Naturalist*, 3: 22-28.

Cameron, J. (1883) *Gaelic names of plants, Scottish and Irish, with notes*. Edinburgh.

Fairweather, B. (1984) *Highland Plant Lore*. Glencoe: Glencoe and North Lorn Folk Museum.

Fife, H. (1985) *The lore of Highland trees: identification, distribution, laws of growth and reproduction, practical benefits, historical significance, symbolism in mythology and custom*. Gartocharn: Famedram Publishers.

Fife, H. (1994) *Warriors and Guardians: Native Highland trees*. Glendaruel, Argyll: Argyll Publishing.

Grieve, M. (1931) *A modern herbal: the medicinal, culinary, cosmetic and economic properties, cultivation and folklore of herbs, grasses, fungi, shrubs and trees with all their modern scientific uses*. London: Jonathan Cape.

Hogg, R. & Johnson, G. W. (1866) *The eatable funguses of Great Britain*. London: Journal of Horticulture & Cottage Gardener Office.

Hooker, W. J. (1821) *Flora Scotica; or, a description of Scottish plants*. London.

Johnson, C. P. (1862) *The useful plants of Great Britain: a treatise upon the principal native vegetables capable of application as food, or in the arts and manufactures*. London: Robert Hardwicke.

Johnston, G. (1853). *The botany of the Eastern Borders*. London: John Van Voorst.

Johnstone, W. G. & Croall, A. (1859-60) *The nature-printed British sea-weeds: a history accompanied by figures and dissections of the Algae of the British Isles*. London: Bradbury & Evans.

McNeill, F. M. (1929) *The Scots kitchen: its traditions and lore, with old-time recipes*. London: Blackie & Son.

Neill, P. (1806) *A tour of some of the islands of Orkney and Shetland*. Edinburgh: Constable.

Pennant, T. (1774) *A tour in Scotland, and voyage to the Hebrides, 1772*. Chester: pr. John Monk, [for the author].

Robertson, J. A. (1869) *The Gaelic topography of Scotland, and what it proves, explained*. Edinburgh.

Sutherland, J. (1683) *Hortus Medicus Edinburgensis*. Edinburgh.

Tuazon-Nartea, J. & Savage, G. (2013) Investigation of Oxalate Levels in Sorrel Plant Parts and Sorrel-Based Products. *Food and Nutrition Sciences* vol 4:838–843.

Vickery, R. (1997) *A Dictionary of Plant-Lore*. Oxford: Oxford Paperbacks.

Withering, W. (1776) *A botanical arrangement of all the vegetables naturally growing in Great Britain*. Birmingham: Cadell & Elmsly.

Left: Coastal plants by Morna Henderson, watercolour (contemporary).

Index to Scientific Names

Acer pseudoplatanus	89, 106, 154
Achillea millefolium	68
Aegopodium podagraria	14, 155
Agaricus campestris	130
Alaria esculenta	35
Alliaria petiolata	91
Allium latifolium	93
Allium ursinum	92
Alnus glutinosa	12, 40, 107
Amanita muscaria	130
Ammophila arenaria	24
Angelica sylvestris	61
Anthriscus sylvestris	160
Arctium minus	156, 159
Argatilis sylvaticus	75
Armeria maritima	30
Artemisia vulgaris	170
Arum maculatum	90, 182
Ascophyllum	33
Aspicilia leprosescens	176
Asplenium scolopendrium	125, 171
Asplenium trichomanes	171
Athyrium	126
Auricularia auricula-juda	131, 151
Bellis perennis	45, 65, 158
Betula nana	12, 108
Betula pendula	108
Betula pubescens	107, 108
Botrychium lunaria	87
Calendula officinalis	169
Calluna vulgaris	136
Caltha palustris	65
Candelariella vitellina	176
Cantharellus cibarius	130
Carduus	173
Chenopodium album	19
Chenopodium bonus-henricus	19
Chondrus crispus	33
Chorda filum	35
Cirsium vulgare	172
Cladonia	184
Cochlearia	26
Conium maculatum	160
Conopodium majus	91
Coprinus comatus	131
Corylus avellana	8, 12, 13, 47, 57, 107, 111
Crataegus monogyna	19, 107, 112, 184
Cytisus scoparius	70
Daucus carota	28
Dianthus barbatus	19
Digitalis purpurea	91, 94
Drymocallis rupestris	135
Dryopteris	126
Dryopteris affinis	124
Dryopteris dilatata	125
Echium marinum	23
Elaeagnus rhamnoides	31
Empetrum nigrum	138
Erica baccifera	139
Erica cinerea	132
Euphrasia	159
Fagus sylvatica	89
Ficaria verna	90
Filipendula ulmaria	42
Fomes fomentarius	130
Frankia	31
Fraxinus excelsior	107, 114
Fritillaria meleagris	66
Fucus	33
Galium aparine	20, 96
Galium odoratum	91
Galium verum	73, 167
Geranium robertianum	19, 91
Geranium sanguineum	91
Geum rivale	38, 47, 65
Geum urbanum	65, 91
Hedera helix	98, 107, 151
Heracleum sphondylium	170
Himanthalia elongata	34, 35
Huperzia selago	141
Hypericum perforatum	17, 100
Hypericum pulchrum	100
Ilex aquifolium	106
Iris pseudacorus	44, 159
Juncus effusus	47
Juniperus communis	142
Laminaria	33
Lathyrus linifolius	74
Lecanora gangaleoides	176
Leymus arenarius	25
Ligusticum scoticum	29, 61
Linum catharticum	169
Lobaria pulmonaria	128
Lonicera periclymenum	91, 151
Lycoperdon	131
Lycopodium clavatum	19, 20, 140

Lycopodium selago	141	*Peltigera*	129, 184	*Prunus spinosa*	117	*Sedum rosea*	29
Lysimachia nemorum	17	*Persicaria*	76	*Pteridium aquilinum*	126, 127	*Senecio jacobaea*	162
		Persicaria amphibia	76			*Senecio vulgaris*	57, 162
Mentha pulegium	105	*Persicaria bistorta*	76	*Quercus petraea*	119	*Silene viscaria*	134
Menyanthes trifoliata	48	*Persicaria hydropiper*	76	*Quercus robur*	107, 118	*Silybum marianium*	173
Mercurialis perennis	91	*Persicaria maculosa*	76	*Quercus species*	12, 89, 107, 118, 136, 147	*Sinapis arvensis*	19
Mertensia maritima	23	*Persicaria vivipara*	76			*Sorbus aucuparia*	106, 107, 121, 131, 150
Meum athamanticum	87	*Petasites hybridus*	65, 94	*Ranunculus acris*	171	*Sphagnum*	39, 62, 133
Mimulus guttatus	64	*Phellinus hippophaeicola*	31	*Ranunculus repens*	171	*Stellaria media*	170
Myrica cerifera	145	*Phytophthora ramorum*	142	*Rhodiola rosea*	29		
Myrica gale	145	*Pimpinella saxifraga*	87	*Rosa canina*	2	*Tanacetum vulgare*	163
Myrrhis odorata	14, 89, 155	*Pinguicula vulgaris*	54, 168	*Rubus idaeus*	104	*Taraxacum officinale*	168
		Pinus sylvestris	12, 89, 106, 133, 146	*Rubus chamaemorus*	143	*Taxus baccata*	107, 122
Narthecium ossifragum	17	*Plantago lanceolata*	171	*Rubus fruticosus*	104	*Tephromela atra*	176
Nasturtium officinale	50	*Polygala vulgaris*	76, 168	*Rumex*	57	*Trifolium*	103, 164
Nymphaea alba	52	*Polygonatum multiflorum*	88	*Rumex acetosa*	82	*Trifolium pratense*	164
		Polystichum	126	*Rumex acetosella*	83	*Trifolium repens*	164
Ochrolechia	129	*Populus tremula*	107, 148	*Rumex aquaticus*	10, 57	*Tussilago farfara*	170
Ochrolechia parella	176	*Porphyra umbilicalis*	33	*Rumex crispus*	57		
Oenanthe crocata	61	*Porpidia albocaerulescens*	176	*Rumex obtusifolius*	57	*Ulex europaeus*	84, 107
Oenanthe lachenalii	61	*Potentilla anserina*	78	*Rumex obtusifolius*	10	*Ulmus glabra*	106
Onopordon aanthium	173	*Potentilla erecta*	80, 136			*Ulva lactuca*	32, 33, 37
Ophioglossum vulgatum	87, 125	*Primula scotica*	86	*Saccharina latissima*	33	*Urtica dioica*	155, 166
		Primula vulgaris	19, 66, 91, 103	*Salix*	12, 57, 58, 107		
Osmunda regalis	94			*Salix fragilis*	58	*Vaccinium myrtillus*	152
Oxalis acetosella	102, 164	*Prunella vulgaris*	87	*Salix herbacea*	58	*Vaccinium vitis-idaea*	153
		Prunus	117	*Salix purpurea*	59	*Viola*	91
Palmaria palmata	33, 36	*Prunus avium*	117	*Salix viminalis*	58		
Parmelia	129	*Prunus padus*	117	*Sambucus nigra*	107, 120, 131, 151	*Xanthoria species*	30
				Scrophularia nodosa	60		

Index to English Names

Adder's tongue 87, 125	Bracket fungus 130	Crowberry 138	Flag 44	Herb Robert 19, 91
Agaricus 130	Brambles 104, 143	Cuckoo flower 47, 66	Flapper bags 156	Hillberry 139
Alder 12, 40, 107	Broad buckler fern 125	Cuckoo sorrel 103	Flesh and blood 80	Hogweed 170
Ash 107, 114	Broad-leaved dock 57	Cuckoo's meat 103	Fleur de lys 44	Holly 106
Aspen 107, 148	Broom 70	Curled dock 57	Fly agaric 130	Honeysuckle 91, 151
	Buckbean 49		Foxglove 91, 94	Horsetails 126
Beech 89	Buckler ferns 126	Daisy 45, 65, 158	Fungi 31, 130	Hot arsmart 76
Bell heather 132	Burdock 156, 159	Dandelion 168		Huckleberry 152
Bilberty 152	Burnet saxifrage 87	Dead men's bells 95	Galls 118	Hundred-leaved grass 68
Birch 107	Butterbur 65, 94	Deal 146	Garlic mustard 91	
Bird cherry 117	Buttercup 171	Docks 57	Golden shield fern 124	Irish moss 33
Bird's nest 28	Butter-plant 54	Dog heather 136	Good King Henry 19	Ivy 98, 107, 151
Bistorts 76	Butterwort 54, 168	Dog lichens 129, 184	Goose grass 79	
Bitter vetch 74		Dog rose 2	Goosegrass 97	Jew's ear fungus 131, 151
Black birch 108	Carrageen 33	Dog tansy 79	Gorse 84, 107	Juniper 142
Black buttons 104	Chanterelles 130	Dog's mercury 91	Ground elder 14, 155	
Blackberries 104	Charlock 19	Downy birch 107, 108	Groundsel 162	Kelps 33, 35
Blackthorn 117	Cherries 117	Dulse 33, 36		King Ellwand's 95
Bladderwrack 33	Chickweed 170	Dwarf birch 12, 108	Hand fucus 36	Knotty birch 108
Blaeberry 152	Cleavers 20, 96		Hart's tongue 125, 171	
Blood-root 80	Cloudberry 143	Elder 107, 120, 131, 151	Hawthorn 19, 107, 112, 184	Lady cakes 103
Bloody bells 95	Clove-root 65	Elm 106, 107	Hazel 8, 12, 13, 47, 57, 107, 111	Lady Mary's thistle 173
Bloody cranesbill 91	Clovers 103, 164	Eyebrights 159		Lady's bedstraw 73, 167
Bloody fingers 95	Club-moss 141		Heath pea 74	Lady's clover 103
Bog asphodel 17	Coltsfoot 170	Fair-days 79	Heather 107, 133, 136, 141	Lady's thimbles 95
Bog mosses 39, 62, 133	Common ragwort 162	Fairy flax 169		Laver 33
Bog myrtle 145	Cotton thistle 173	Fairy's thimbles 95	Heather berry 139	Lesser celandine 90
Bogbean 48	Cow parsley 160	Fat hen 19	He-heather 136	Lichens 30, 128, 184
Bog-nut 49	Cowberry 153	Ferns 125	Hemlock 160	Ling Heather 136
Bracken 126, 127	Crottle 129	Field mushroom 130	Herb bennet 91	Lingonberry 153
		Figwort 60		Lint 169

180

Lords and ladies 90, 182	Pendunculate oak 107, 118	Scotch thistle 173	Sour clover 103	Tree lungwort 128
Lyme grass 25		Scot's pine 12, 89, 106, 133, 146	Sour dock 83	Tuberous pea/ Tuberous vetch 74
	Pennyroyal 105		Spear thistle 172	
Maidenhair spleenwort 171	Pignut 91	Scot's primrose 86	Speedwell 66	Vine 107
Marigold 169	Pig's roots 79	Scottish dock 57	Spignel 87	Violets 91
Marram grass 24	Poor man's lettuce 103	Scurvy-grass 26	Spotted arsmart 76	
Marsh marigold 65	Primrose 19, 66, 91, 103	Sea buckthorn 31	Spotted knotweed 76	Water avens 38, 47, 65
Marsh trefoil 49	Puffballs 131	Sea lettuce 32, 33, 37	St John's wort 17, 100	Watercress 50
Mascorn 79	Purple moor grass 87	Sea parsley 29	Stag's-horn club-moss 19, 20, 140	Water-dropworts 61
Meadow buttercup 171		Sea tangle 32		Water-pepper 76
Meadow queen 42	Ragweed 57, 162	Seaweeds 32, 94	Sticky catchfly 134	Whin-cow/Whins 85, 107
Meadowsweet 42	Ragwort 162	Sedges 44	Stinking Billy 19	White clover 164
Meduart / Medwort 42	Ramsons 92	Self-heal 87	Stinking Bob 19	White water-lily 52
Mermaid's tresses 35	Raspberries 104	Sessile oak 119	Stinking Roger 60	Whortleberry 152
Midden mylies 19	Red clover 164	Shaggy ink-cap 131	Sugar-wrack 33	Wild Angelica 61
Midden weed 19	Red shank 83	Shamrock 103	Sutherland kale 6	Wild carrot 28
Milk thistle 173	Redshank 76	Sheep's sorrell 83	Sweet cicely 14, 89, 155	Wild iris 44
Milkwort 76, 168	Ribwort plantain 171	Shepherd's knot 80	Sweet William 19	Wild leek 92
Monkeyflower 64	Rock cinquefoil 135	Shetland monkeyflower 64	Sweet woodruff 91	Wild mercury 91, 95
Moonwort 87	Rose-noble 60	Shield ferns 126	Swine beads/ Swine's grass 79	Willows 12, 57, 58, 107
Moor grass 79	Roseroot 29	Silver birch 108	Sycamore 89, 106, 154	Witch's paps 95
Moss crop/Moss grass 79	Rowan 106, 107, 121, 131, 150	Silverweed 78		Wood avens 65, 91
Mountain ash 151		Sloes 117	Tangle 33	Wood sorrel 102, 164
Mugwort 170	Royal fern 94	Small bistort 76	Tansy 163	Wracks 33
	Rush 47	Snakeroot 76	Thistle 173	
Nettle 155, 166		Snakeshead fritillary 66	Thongweed 34, 35	Yarrow 68
	Scot's lovage 29, 61	Snakeweed 76	Thousand-leaved clover 68	Yellow Flag Iris 44, 159
Oaks 12, 89, 107, 118, 136, 147	Scotch mahogany 40	Soft rush 47	Thrift 30	Yellow pimpernel 17
	Scotch mercury 91, 95	Solomon's seal 88	Tod's tails 95	Yew 107, 122
Oyster plant 22	Scotch parsley 29	Sorrel 82	Tormentil 80, 136	

181

Index to Gaelic Names

Achlasan Chaluim Chille	17, 101
Airgean	34
Aiteann	142
Aithair an duileasg	34
Am boinne fola	76
An raineach mhòr	127
Beachnuadh boireann	100
Bealaidh	70
Bearnan Bride	168
Beatha	108
Beatha beag	108
Beith charraigeach	108
Beith dubach	108
Beithe dhubh-chasach	108
Biolair uisge	51
Brisgean	79
Buaghallan	162
Cairt leamhna	74
Cairt-láir	80
Calltainn	111
Caora-mhiteag	152
Caorann	151
Carraceen	34
Carran	27
Coinneach dearg	63
Conasg	85
Copag albannach	57
Copag chamagach	57
Copag leathann	57
Creamh	92
Creas Chu Chulain	42
Critheann	148
Curran fiadhain	28
Darach	118
Darach Gasagach	118
Deanntag	166
Driamlach	35
Dris	104
Droman	120
Duileasg	36
Duilleag-bhàite bhàn	52
Eadha	148
Earr thalmainn	68
Feada coille	103
Feàrna	40
Fiodhag	117
Fionnlach	63
Fraoch	136
Geanais	117
Gille mu lean	35
Giuthas	146
Glùineach theth	76
Gort	98
Huath	107
Iogh	107
Ioghar	123
Iteodha	160
Lubhar	122
Langadair	35
Leabadh ban-sith	73
Leadan liosda	156
Luachair bhog	47
Luis	107
Lus a madhaidh-ruadh	140
Lus an leasaich	73
Lus Chaluim Chille	17, 100
Lus chrann ceusaidh	76
Lus na caithimh	91
Lus na fala	101
Lus na feannaig	138
Lus nam Ban-sith	94
Lus nan oighreag	143
Lusan-glùineach	76
Lus-nan cnapan	60
Lus-nan Laogh	29
Mathair an duileasg	34
Mointeach liath	63
Mòthan	54
Muin	107
Muran	24
Neoinean	159
Nuin	107
Oir	107
Onn	107
Peith bhog	107
Preas nan airneag	117
Roid	145
Ruis	107
Samh	83
Sealbhag nan Caorach	83
Seamrag Bhan	164
Seamrag Dhearg	164
Seileach	58
Seileasdair	44
Seud Chaluim Chille	17
Sgitheach	112
Shunnis	29
Sùbh-craoibh	104
Suil	107
Sunais	29
Teine	107
Torranan	60
Tri-bhileach	49
Uinnseann	114
Ur	107

Acknowledgements

Written by Gregory J. Kenicer

Design, layout and photo retouching by Caroline Muir

Production, editorial and project management by Donna Cole and Lorna Mitchell

Illustration coordination by Sarah Roberts, and Jacqui Pestell MBE

Photography by Lynsey Wilson

This work was inspired by the illustrations produced by many members of the Scottish botanical art community, and we are extremely grateful for their support and encouragement throughout the process.

Particular thanks are due to the members of the Scottish Botanical Art Collective and to Jenny Haslimeier, Kim Howell, Phil Lusby, Catherine Conway-Payne, Claire Banks, Lizzie Sanders, Alice Young, Elinor Gallant, Ian Edwards, Leonie Alexander and the Royal College of Physicians.

The artists are:

Carolann Alexander	Janette W Dobson	Jan Miller	Sandra Russell
Janis Bain	Janet Dyer	Hazel Morris	Lizzie Sanders
Claire Banks	Sheila Anderson Hardy	Kathy Munro	Alexa Scott Plummer
Marie Barbour	Jenny Haslimeier	Gloria Newlan	Elisabeth Scott
Victoria Braithwaite	Marianne Hazlewood	Mary O'Neil	Leigh Tindale
Geoffrey Brown	Morna Henderson	Jacqui Pestell	Sharon Tingey
Lyn Campbell	Sarah Howard	Kathy Pickles	Fran Thomas
Eleanor Christopher	Jan Kerr	Julie Price	Judith La Trobe
Claire Dalby RWS RE	Charlotte Long	Coral Prosser	Margaret Walty
Anne Dana	Nicola Macartney	Jocelyn Anne Rabbitts	Fiona Ward
Brenda Davies	Nichola McCourty	Sarah Roberts	Janet Watson
	Clare McGhee	Linda Russell	

With special thanks to Sir Charles and Lady Fraser

Lichens by Margaret Walty, acrylics (contemporary).

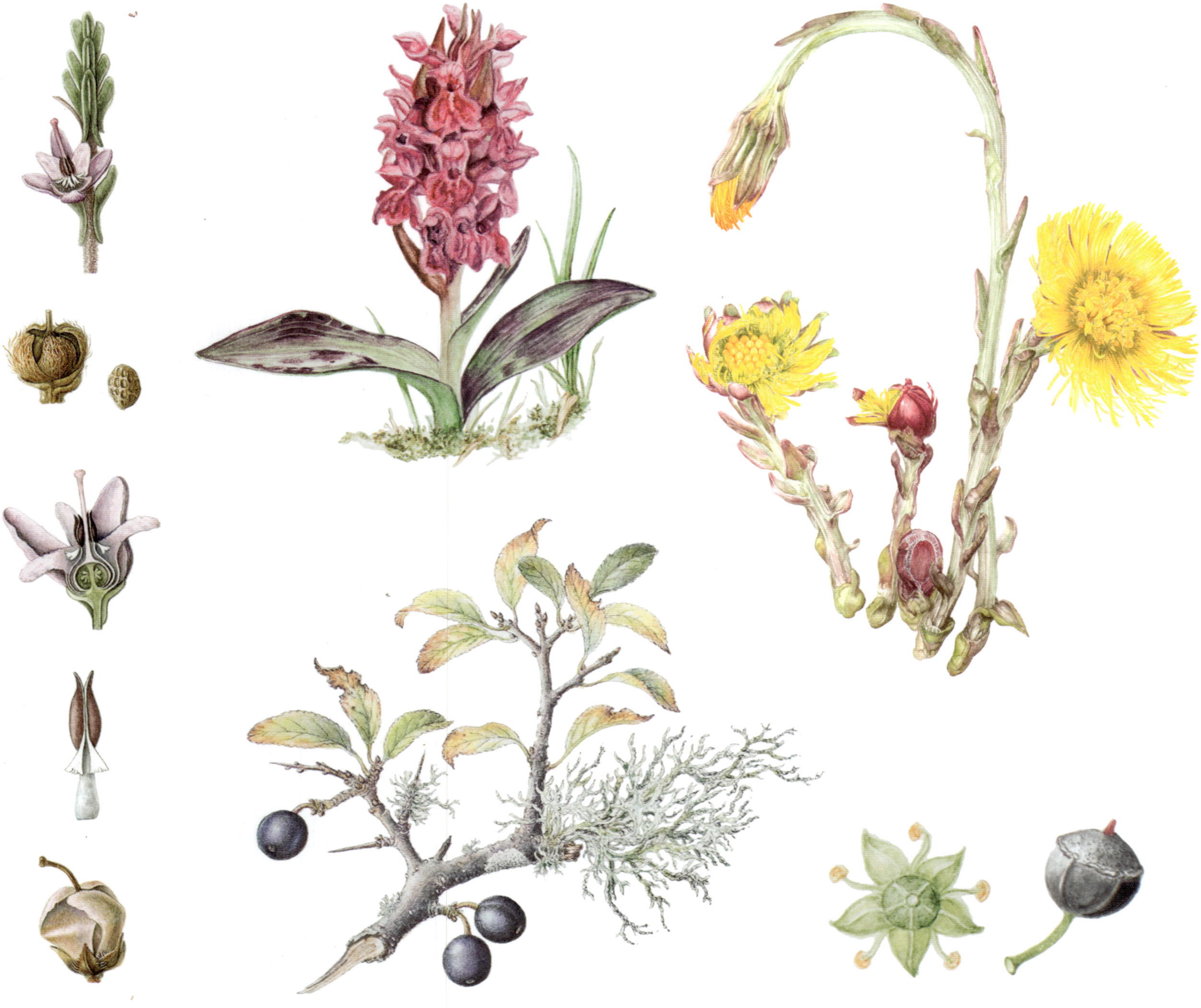